IMAGES
of America

FIREFIGHTING IN HOPKINSVILLE

FIRE STATION SHIELD. The great shield of the Hopkinsville Fire Department is portrayed in this handsome sign attached to the front wall of the West First Street Central Fire Station. It includes the seal for the city of Hopkinsville, founded in 1797, which was adopted by the city council on April 27, 1872. This seal includes a rendering of the pennyroyal plant. The name pennyroyal, a plant in the mint family, was applied to this geographic region of Kentucky when the state was divided in 1853. The shield features a fireplug and ladders along with the department motto, "Save Lives—Protect Property." (Courtesy of Christian County Historical Society.)

ON THE COVER: The most sensational fire in Hopkinsville history occurred when the 93-room, 4-story Hotel Latham burned to the ground on August 4, 1940. The Sunday afternoon event attracted a large crowd, and many images survive, along with a color home-movie film. Hopkinsville's 1927 Chevrolet pumper was hard at work in the intersection of Eighth and Virginia Streets. (Courtesy of Christian County Historical Society.)

IMAGES
of America

FIREFIGHTING IN HOPKINSVILLE

Chris Gilkey and William T. Turner
Foreword by Chief Fagan W. Pace Jr.

ARCADIA
PUBLISHING

ISBN 978-1-5316-3346-2

Published by Arcadia Publishing
Charleston SC, Chicago IL, Portsmouth NH, San Francisco CA

Library of Congress Catalog Card Number: 2007934165

For all general information contact Arcadia Publishing at:
Telephone 843-853-2070
Fax 843-853-0044
E-mail sales@arcadiapublishing.com
For customer service and orders:
Toll-Free 1-888-313-2665

Visit us on the Internet at www.arcadiapublishing.com

Dedicated to our firefighters, past, present, and future, for their heroism, dedication, commitment, courage, and honor.

CONTENTS

Acknowledgments 6

Foreword by Fire Chief Fagan Pace 7

Introduction 8

1. Fire Stations 9
2. Firemen 25
3. Chiefs 37
4. Equipment 45
5. Training 63
6. Fire Prevention 71
7. Landmark Fires 81
8. Hotel Latham Fire 99
9. Rural Volunteer Fire Departments 113

Acknowledgments

We would like to express appreciation to the following individuals for their assistance and motivation in compiling this book: LaDonna Dixon Anderson, Patsy Bell, Greg Bequette, Eric Bleidt, Lewis Brooks, Edgar Cayce, Cecelia Cloos, Rick Denny, Glenn Douglas, James Douglas, Joe Dorris, Jim Eli, Beth Lilly Everett, Neil Farmer, Elsie Rex Foster, Lindsey Gilkey, James E. "Randy" Graham, Celestine Graves, George Hampton, Carla Hill, Dyke Higgins, Henry Jako, Lyn Cate James, *Kentucky New Era*, W. E. Knox, James "Skip" Latham, L. D. Martin, Jada Mason, Kevin McDowell, Carl McGee, Lynn Moe, Robert Moody, James H. Moore, Nancy Moore, Cathy O'Steen, Bobby Outland, Hopkinsville fire chief Fagan W. Pace Jr., Keith Sholar, David C. Smith, Emily B. Perry, Roy Stadleman, Annie Pierce Steger, Doris Stokes, LaVena E. Turner, Walter "Buster" Wallace, Lillian Randle Wilkins, John W. "Woody" Winfree, Ben S. Wood III, and Wayne Yates. All of the images are courtesy of the Christian County Historical Society.

FOREWORD

It is a rare and distinct honor to be asked to foreword a publication of this caliber. William Turner and Chris Gilkey have put together an extremely informative, entertaining, and comprehensive look at the first 175 years of firefighting in Hopkinsville. An overview of the present state of the Hopkinsville Fire Department is an appropriate place to begin.

Current authorized staffing consists of 80 uniformed and one civilian employee. These employees serve Hopkinsville's current approximate population of 30,000 citizens within the city's 30.7 square miles, 24 hours a day, 7 days a week, 365 days a year out of four stations: Station 1 at 116 West First Street; Station 2 at 1108 Skyline Drive; Station 3 at 3203 Canton Pike; and Station 4 at 9505 Eagle Way Bypass.

From a statistical perspective, our staffing equates to 2.60 firefighters per 1,000 citizens and, coincidentally, 2.60 firefighters per square mile. Fiscal year 2006–2007's approved fire department budget is $5,185,148. Additional demands continue to accrue from new building and annexation. In the 10-year period between 1996 and 2006, there was an estimated $500 million in new construction building permits issued in Hopkinsville.

In the calendar year 2006, the Hopkinsville Fire Department responded to 1,207 calls for emergency service. These include structure fires, vehicle fires, EMS assists, hazardous material response, miscellaneous rescue, and so forth. Additionally, the department responded to 4,162 calls for emergency medical treatment and transport.

Over the years, the Hopkinsville Fire Department's capabilities and responsibilities have greatly expanded. These capabilities currently include but are not limited to the following: fire suppression, fire prevention, fire inspection, arson, investigation, fire education, building inspection, Level A hazardous material response, advanced life support emergency medical services, auto extrication, heavy rescue, high angle rescue, confined space rescue, and swift water rescue. The fire department rolling stock is currently quite comprehensive. Obviously, specialized and highly technical training and equipment have expanded as well to adequately support critical fire, EMS, and rescue disciplines.

While capabilities and equipment have grown and improved throughout the history of the department, the core values of loyalty, dedication, courage, and service to the community remain the cornerstones of the Hopkinsville Fire Department. The department is very proud of its past and present. We look confidently to the future and optimistically embrace the challenges and the rewards of proudly serving our fellow citizens.

—Fire Chief Fagan W. Pace Jr.

INTRODUCTION

The Hopkinsville Union Fire Company, organized March 10, 1832, represented the first attempt to provide fire protection for Hopkinsville, then with a population of 1,260. It was followed by the creation of succeeding volunteer groups, called "runners," through the remainder of the 19th century. On March 23, 1901, a paid professional department was established, with the firemen and equipment housed in the city hall and town market house at the southwest corner of Fifth and Main Streets. Four years later, the new Central Fire Station, located on East Ninth Street between Liberty and Clay Streets, was completed. On October 3 and 4, 1924, this building was destroyed by a fire caused by faulty wiring. A new station was built within the original walls of the burned building. This facility served as the Hopkinsville Fire Station until the department moved to the new central station on West First Street in May 1964. Four substations have been added to the force: Fort Campbell Boulevard, Canton Pike, Skyline Drive, and the Eagle Way Bypass. The Boulevard station was later closed, and three substations serve the city today along with the main station. The old East Ninth Street station is currently undergoing restoration for development into the "Woody" Winfree Fire-Transportation Museum.

Throughout the years, Hopkinsville firefighters have maintained a special social status in the minds of local citizens. The firefighters are always there to help in a host of crises including fires, accidents, heath problems, floods, and even the occasional rescue of animals. The force has conducted an annual toy drive, including repair, to provide a more joyful Christmas for those less fortunate. They also have sold Christmas trees to raise money for the underprivileged. Beginning after World War II, the department has led an annual fire prevention week in October. Fire prevention programs are presented to area students. Public demonstrations of firefighter training are performed for local residents in the downtown business section. All firefighters engage in weekly training.

This book includes a pictorial review of the eight buildings, past and present, used to house the department. A steam pumper and horse-drawn wagons of the 19th century and motorized trucks of the 20th and 21st centuries are featured. Group pictures of firefighters from the 1890s through today feature the integration of African Americans and women on the force. Individual pictures of fire chiefs from the 1890s are included. The three training towers—utility pole, concrete block, and brick—are shown. The two town clocks are illustrated. Numerous fires, including at the Hotel Latham, are also included. The commitment of 14 county volunteer fire departments portrays the willingness of rural people to go above and beyond the call of duty. Featured are ruins of the fire-destroyed station, terrific action shots of the Hotel Latham fire in 1940, construction of substations, and images taken during Fire Prevention Week.

One

Fire Stations

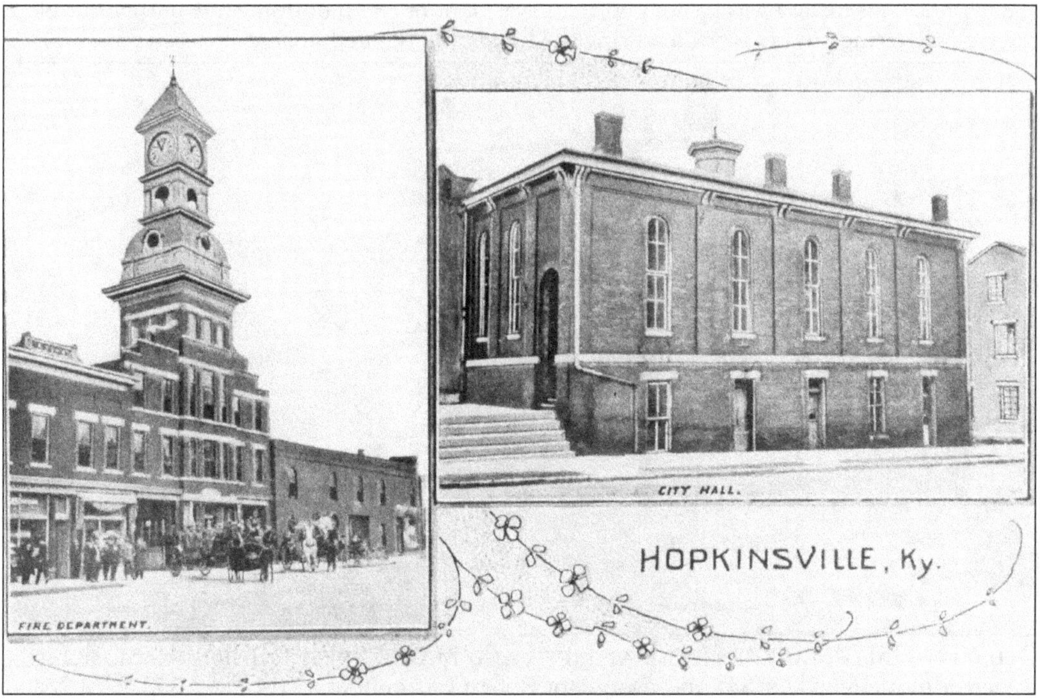

Central Fire Station and City Hall. S. H. Kress 10¢ Store published and distributed this color postcard featuring the fire station and city hall about 1915. City hall, located on the southwest corner of Fifth and Main Streets, was constructed in 1880–1881 and, at the time pictured, housed the police station on the lower floor with the city courtroom above.

FIRE STATION AT FIFTH AND MAIN STREETS. Longtime firemen Charlie Hord and John Lawson pose in front of the Hopkinsville Fire Department's second location, at Fifth and Main Streets, about 1902. This facility was located on the lower floor of city hall from 1881 until 1905. Note the fancy lettering on the doors, featuring the hands and pointed fingers.

OLD CITY HALL, MARKET HOUSE AT FIFTH AND MAIN STREETS. In 1880–1881, architect Robert Mills designed for $35 and contractor Robert Orr built at a cost of $2,800 at Fifth and Main Streets the new city courtroom and engine house, as the fire station was first known. From 1871 until 1881, the department occupied the old market house, pictured at right. The market house was converted into the city jail and the police department occupied the lower floor of city hall after the fire station moved to East Ninth Street.

CENTRAL FIRE STATION. The City of Hopkinsville constructed the East Ninth Street Central Fire Station in 1904–1905. W. A. Long served as architect, with Forbes Manufacturing Company building the structure, which was located on the north side of East Ninth Street between Liberty and Clay Streets. The 85-foot tower, used for drying fire hoses, contained a town clock manufactured by the E. Howard Clock Company of Boston, Massachusetts. The four clock faces, featuring Roman numerals, were 7 feet in diameter. The town fire bell, weighing 1,018 pounds and cast by the Henry McShane Foundry, in Baltimore, Maryland, was bought by the city for $225 on September 6, 1887. The bell hung in the courthouse cupola, 60 feet above the ground, from October 1, 1887, until March 13, 1905, when it was installed in the East Ninth Street station. The bell served a triple purpose: it sounded the fire alarm, struck the hour, and notified volunteer firemen, by number of rings, the district in which the fire was located. This picture was made on May 28, 1905, soon after the new station was occupied.

CENTRAL FIRE STATION, 1920. Firemen watched the town go by from their vantage point at the East Ninth Street Central Fire Station around 1920 in split-bottom chairs along the sidewalk. Motorized fire trucks had been added the year before, and the fire horses had been retired to pasture. The fire department had grown from 11 men in 1905 to an even dozen along with the addition of several runners. These men worked at regular jobs, slept at the fire station, and served on a standby basis in case of a large fire. At 11:45 p.m., October 3, 1924, the station was discovered to be on fire! Faulty wiring in the bathroom ceiling of the family living quarters set the stage for the destruction of this landmark. Firemen on duty John Lawson, Charles Dulin, Ernest Hayden, Joe East, Guy Smith, and Otto and Joe Fish were roused from their sleep in the dormitory on the west side of the second floor. The two fire trucks were removed, and a valiant effort was made to save the structure. The building burned down to the second floor, and the stately town clock tower collapsed as the fire bell was striking midnight.

NINTH STREET CENTRAL FIRE STATION, 1938. This *c.* 1938 aerial image of the Ninth Street station captures some festive occasion as bunting is draped from the second-floor fire station windows. This station was rebuilt within the walls of the 1904 building and was completed January 12, 1925. John T. Waller was the architect, and the contractor was Ellis J. Roper. The reconstruction cost $6,149.

THE OLD TOWN CLOCK, 1964. The construction firm of Long and Whoberry was given the contract to rebuild the town clock tower in September 1926 for $4,289, and it was completed in December. Constructed of metal, it has withstood sun, wind, and rain for 80 years. This image was made May 17, 1964, on the day the fire department moved from this landmark to the new West First Street central station. Note the fire bell in the third clearstory window from left.

CLOCKWORKS, SEPTEMBER 2007. The Women's Club of Hopkinsville announced in February 1926 plans to install a new city timepiece. The fund drive raised $2,000 in six months, and the clockworks were installed in the new tower on January 17–24, 1927.

THE PENDULUM. The driving force for the operation of the large timepiece alternates between this 75-pound pendulum and about 375 pounds of brick-shaped iron positioned on a platform. The pendulum box is positioned a full floor below the clockworks. The driving weight is located in a vertical box. Aluminum hands 42 and 32 inches long indicate the minute and hour.

Old Fire Bell. The 1887 fire bell, which survived the 1924 fire though it fell some 50 feet, is still in operation today. It was reinstalled in the new town clock tower in late 1926. The striking mechanism consists of an exterior hammer that strikes the outside of the bell about one third of the way up the lip. The brick-shaped irons, weighing 1,500 pounds, drive the striker.

Clock Hand Gears. Twenty-five years after installation, the clock took an automatic rest. It was repaired and put back in operation in 1953. It remained silent for 19 years after the fire station moved. In 1983, the town clock tower was repainted and the clock put back into operation. The tower was painted in 1993, and in 2007, in addition to repainting, the clockworks were removed, repaired, cleaned, and put back into operation at a cost of $8,000, a gift from the Hopkinsville High School classes of 1937 and 1938. The clock hand gears are featured here.

FUTURE FIREMEN ON THE POLE. About 1952, this group of Cub Scouts experimented with sliding down the brass pole at the East Ninth Street central station. This event was a part of an extensive annual program in fire prevention. Several Scout groups toured the station in the late 1940s and throughout the 1950s as a part of Fire Prevention Week.

HERE COMES ANOTHER FUTURE FIREMAN. In July 1953, a group of Cub Scouts experiences a tour of the Hopkinsville Fire Station. The brass pole provided the opportunity every school boy wanted. Note the firemen's gear hooked on the wall, the large city map, several calendars, and the pipe from the coal-fired stove that heated the lower floor, where the fire trucks were located.

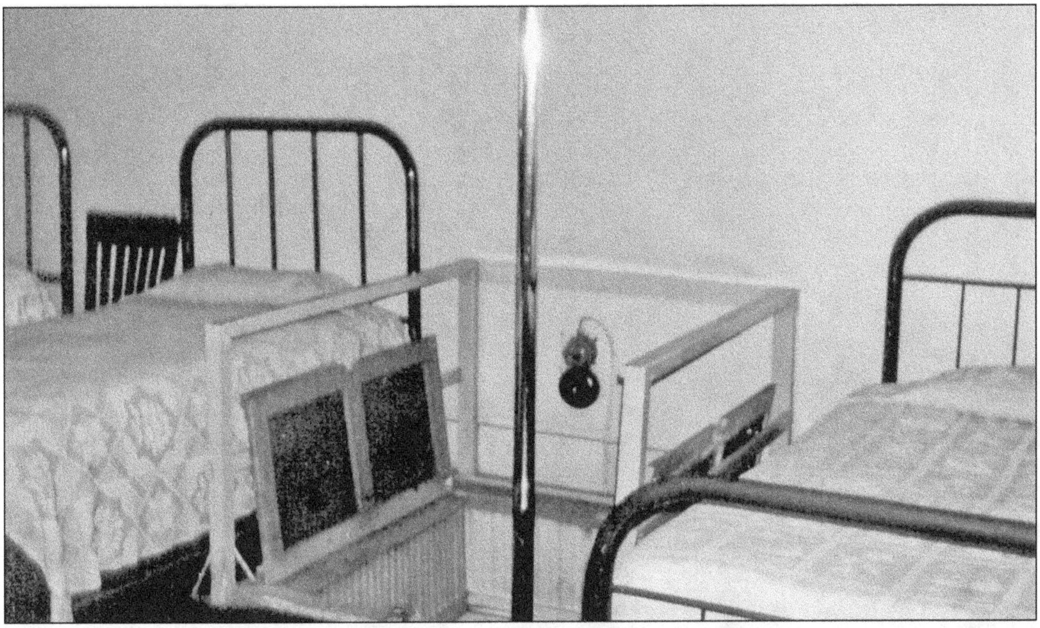

FIREMEN'S QUARTERS AND THE BRASS POLE. About July 1959, fireman Jim Moore photographed a section of the sleeping quarters and the brass pole leading to the first floor. The pole, a necessity in two-story fire stations, became a tradition throughout the country. This pole survives and will be incorporated in the restoration of the East Ninth Street central station, to be known as the Woody Winfree Fire-Transportation Museum.

FIREMAN'S CLOSET. For the sake of history, fireman Jim Moore had the presence of mind to photograph his closet about July 1959. There were 29 firemen on the force at that time, and each one had his own storage space for personal effects. A calendar, a mirror, a towel rack, and clothes on hangers are included with a bottle of Old Spice on the shelf above.

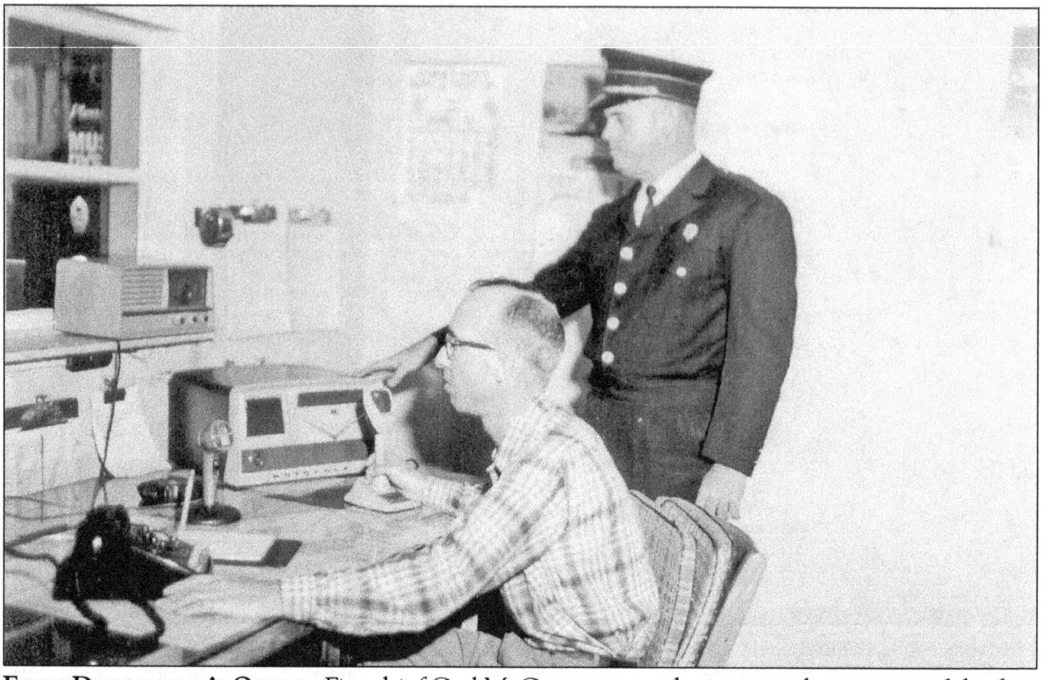

First Dispatcher's Office. Fire chief Carl McGee oversees the inaugural operation of the first two-way radio system in the dispatcher's office at the East Ninth Street central station. This 1962 milestone of progress was first operated by the unidentified disabled veteran. The Emergency Operation Center represents a great leap forward in the past 45 years.

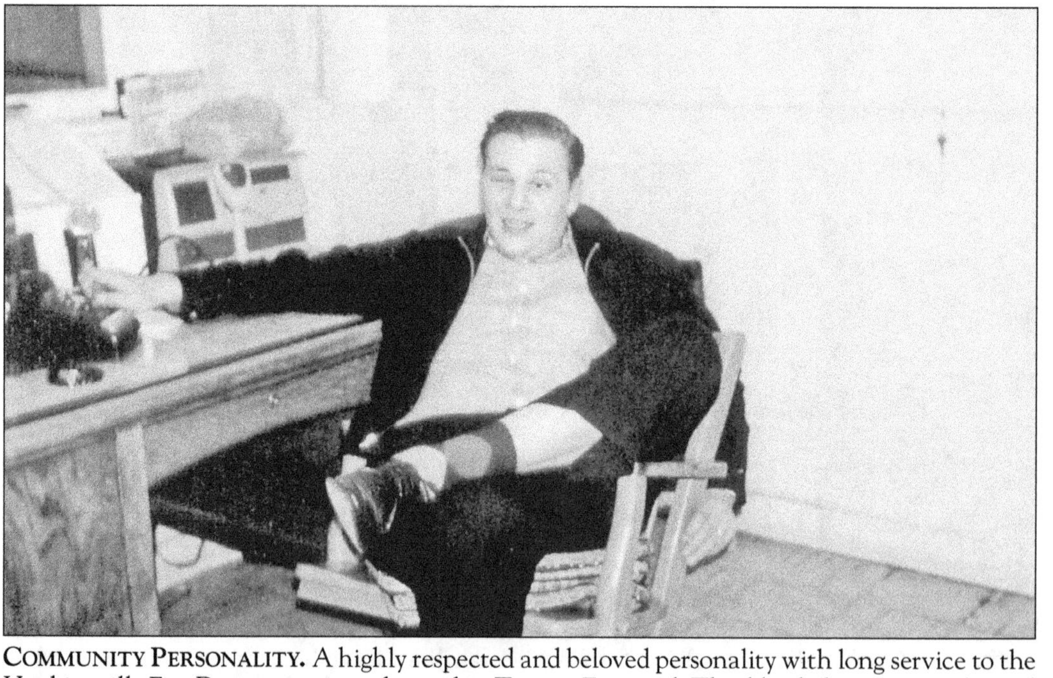

Community Personality. A highly respected and beloved personality with long service to the Hopkinsville Fire Department was dispatcher Tommy Frommel. This blinded veteran performed his duties as dispatcher for many years before his death in 2007. He is pictured in the dispatcher's office at the East Ninth Street central station.

THE LAST RUN. The last run from the East Ninth Street central station was made on May 17, 1964, when the department moved to the new West First Street central station. Fire chief Carl McGee stands at right beside the cab of the 1960 American LaFrance aerial truck. The fire bell from the last horse-drawn wagon is shown in the clearstory window, third from left.

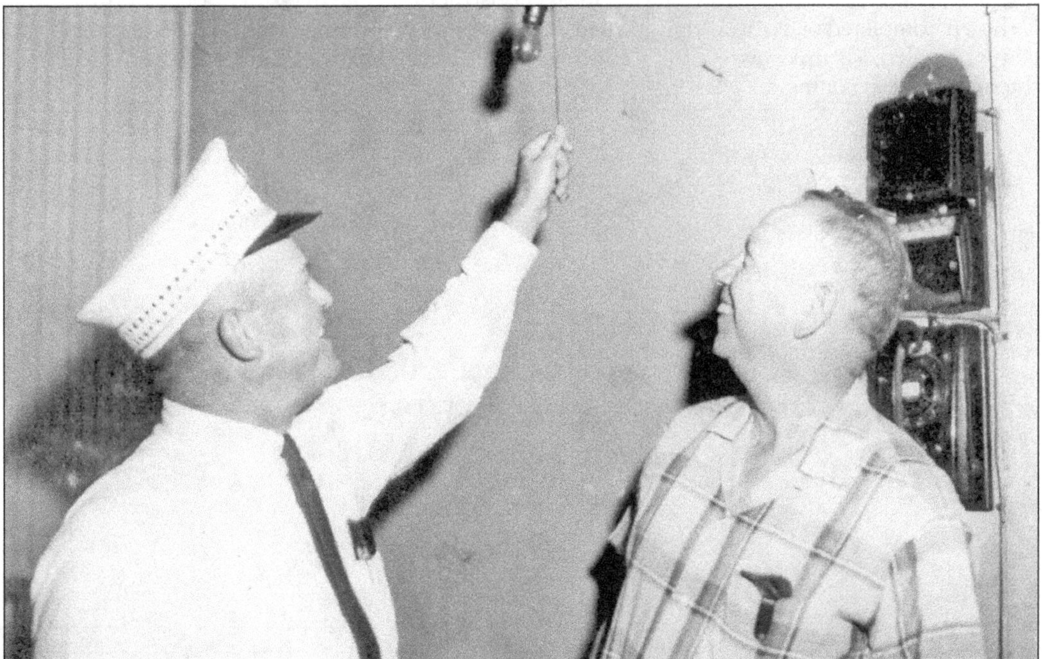

TURING OUT THE LIGHT. Fire chief Carl McGee turns out the light over the call phone at the East Ninth Street central station. Assistant chief Walter "Buster" Wallace looks on as the end of an era arrives. The light bulb had been on day and night for over 40 years. The light bulb, along with the brass fireman's pole, was taken to the new main station on May 17, 1964.

CENTRAL FIRE STATION UNDER CONSTRUCTION. This September 26, 1963, photograph shows the new West First Street central station under construction. The five-story brick training tower is shown completed at right. Original plans called for a two-story station; the reason for it not being completed is unknown. This station was built in the curve of Little River on the site of the original town dump.

WEST FIRST STREET CENTRAL STATION. The West First Street central station is featured in this October 1969 photograph, along with the firemen's training tower, five years after moving to this location. The station and the training tower were designed in 1963–1964 by Lee Potter Smith and Associates. Rader Construction Company was the contractor. Note the old siren located on the roof in the era before the new Emergency Operations Center was constructed on the east end of the station.

EMERGENCY OPERATIONS CENTER. In 1993, city and county emergency services were merged into the Emergency Operations Center (EOC) with 911 communications capabilities. The EOC occupies an addition to the West First Street central station. Here dispatcher Lucas Stagner works at one of the dispatcher stations.

SAFETY COMPLEX. An aerial view made on April 7, 1976, provides an interesting view of the emergency services facilities of Hopkinsville. West First Street extends in front of the police station at right, with the EOC attached to the Central Station at left. The brick training tower stands on the bank of the north or town fork of Little River. This area was developed as a part of the Dr. Frank H. Bassett Urban Renewal Project in the early 1960s.

Under Construction. Construction was under way in early 1960 in this image, which portrays the first new fire department substation. On December 8, 1959, ground was broken for Substation No. 2, a two-bay, one-story brick fire station. It was completed in 1960 and was located on Fort Campbell Boulevard across from the Dr. Pepper Bottling Company. It was closed in 1975.

Work Horses. The old and the new, the 1940 Dodge pumper and the 1964 American LaFrance, sit proudly side by side in front of Substation No. 2 on Fort Campbell Boulevard. This station was in use for 15 years, until it was replaced in 1975 by the new Substation No. 2, located on Skyline Drive.

ANOTHER SUBSTATION. Substation No. 3 was constructed in 1963 and opened in 1964 on Canton Pike at Country Club Lane. Tom M. Paine was the architect for this project, and the contractor was Rader Construction Company. This image reveals the 1947 Seagrave in the left bay and the 1960 Seagrave in the right bay. This substation, still in operation, serves the west side of Hopkinsville.

SKYLINE DRIVE. The new Substation No. 2, located off Skyline Drive, is featured in this 2007 photograph. Tom M. Paine was the architect for this substation, and Rader Construction was the general contractor. Opened in 1975, this station serves the industrial southeast side of Hopkinsville.

EAGLE WAY BYPASS. The newest Hopkinsville fire facility is Substation No. 4, located on the Eagle Way Bypass near the Stadium of Champions. Rader Construction Company was the contractor for the newest substation. It was completed and opened in 2000. This station represents a new architectural style for Hopkinsville fire substations.

A FRESH COAT OF PAINT. This August 2007 view of the town clock, long the symbol of the Hopkinsville–Christian County Chamber of Commerce, indicates that a fresh coat of red and white paint can give the old town landmark a refreshing look. Painter Danny Wells uses a bucket lift attached to a crane provided by General Steel Corporation as a complimentary service to the community to paint the town clock on a very hot and windy day.

Two

FIREMEN

LAST VOLUNTEERS AND FIRST PAID FIREMEN. This image of firemen at the old Fifth and Main Street Station shows the last volunteers, part of the Peerless Fire Company, reorganized January 1, 1901. They, like their predecessors throughout most of the 19th century, served the public as fire volunteers. When the Hopkinsville Fire Department was organized on March 23, 1901, the group pictured became the first paid force in Hopkinsville. The group includes, from left to right, Chief George E. Randle, Ellis H. Roper, George W. Thacker, H. M. Harrison, Ed H. Hester, Ed P. Fears, James Western, E. Polk Renshaw, and John L. Lawson. The teamster seated on the wagon behind the men is Elza H. Cofer. The volunteers resided in their individual homes, with a sleeper staying at the station after about 1889. With the organization of the paid department, sleeping rooms were provided in the old market house directly behind the fire station. The men worked in shifts, the structure of which has been changed several times in the past century.

LINE UP AT THE NEW CENTRAL FIRE STATION. On the morning of May 28, 1905, members of the Hopkinsville Fire Department stood to have this image made in front of the new East Ninth Street central station. The men are, from left to right, (first row) Charlie Hord, William Turner, Ed Hester, E. P. Fears, Chief George E. Randle, John Thompson, Ellis Roper, and John Lawson; (second row) Ed Schmidt, Bob Tunks, Ed Marshall, and Welburn Guthrie.

FIREMEN, 1932. Members of the Hopkinsville Fire Department in this image taken over 70 years ago include, from left to right, (first row) Garland Morse, Ed East, Charles Dulin, John Payne, and John Lawson; (second row) Stanley Ward, Pete Perry, Chief Joe East, Austin Morgan, and Ernest Hayden.

FIREMEN, 1944. The Hopkinsville Fire Department during World War II consisted of, from left to right, Chief Joe East, Ed East, Stanley Ward, Carl McGee, Buster Wallace, Leland Davis, Bill Griffin, Doc Dulin, and Pete Perry. The personnel turnover during the 1930s and 1940s was very limited.

FIREMEN, 1950. A half century ago, after years with a limited number of firemen, the force began to grow. The department included, from left to right, Joe Ward, Buster Wallace, Ed East, Carl McGee, Forgey Mobley, Doris James, Leland Davis, Bill Griffin, Haywood Scruggs, Stanley Ward, Dennis Blandford, Pete Perry, building inspector Clarence King, and Chief Joe East.

Firemen, 1952. With the 1947 Seagrave pumper as a backdrop, the department includes, from left to right, (first row) building inspector Clarence King, Lawrence Hinton, Floyd Cannon, Bill Griffin, Stanley Ward, Carl McGee, Albert Lacy, Chief Doris James, Pete Perry, Dennis Blandford, Dillon Berkley, Norman Cowan, and Noble Warfield; (second row) Leland Davis, Joe Ward, Buster Wallace, and Haywood Scruggs.

Firemen, 1963. This image is the last group picture made at the East Ninth Street central station. Included in the group are, from left to right, (first row) Norman Cowan, Leland Davis, Dennis Blandford, Jim Moore, dispatcher Tom Frommel, Ray Bryant, Harry Mason, John Adams, and James Nichols; (second row) Harvey Shemwell, James Shemwell, Buster Wallace, Eugene Powell, Pete Perry, Ronald Fitzgerald, Floyd Cannon, Noble Warfield, and Albert Lacy; (third row) Hugh Dossett, L. G. Davis, Howard Siggers, Ronnie Blakeley, Dillon Berkley, Joe Cravens, Bill Griffin, Jessie Lancaster, and Chief Carl McGee.

FIREMEN, 1965. From left to right are (first row) Walter Pruitt, James Shemwell, Elvis Powell, Mickey Petty, Harry Mason, Larry James, Red Adams, James Hairfield, and Walter Edwards; (second row) Chief Carl McGee, Buster Wallace, Bill Griffin, Norman Cowan, L. E. Davis, Jim Moore, James Nichols, Noble Warfield, and Albert Lacy; (third row) Jessie Lancaster, Ronnie Fitzgerald, Barkley Gilliland, Tommy Shepherd, Dennis Blandford, Jeff Ezell, Frank Adams, Don McGowan, Tom Frommel, and Harvey Shemwell; (fourth row) Dillon Berkley, Earl Saltsman, Tom House, Hugh Dossett, John Sugg, Richard Lancaster, Coy Siggers, Mutt Petty, and Tom Knight.

FIREMEN, 1968. From left to right are (first row) Tommy Shepherd, James Petty, Larry James, John Curtis, Howard Lee, Bill Combs, James Flick, and Raymond Roberson; (second row) Raleigh Garrett, James Hairfield, Chief Floyd Cannon, Red Adams, Hugh Dossett, Jim Moore, Harry Mason, Harvey Shemwell, Earl Craft, and Norman Cowan; (third row) James Cannon, John Brown, John Kershaw, Richard Lancaster, Joe Cravens, Dillon Berkley, Earnest Craft, Larry Lee, Bill Stagner, Mack Moore, unidentifed, Kenneth Hancock, Mike Ellis, and Jerry Craft.

NEW MASCOT. In the fall of 1976, city firemen named Blackie their official mascot. The pooch had wandered into the Central Fire Station on West First Street six months before and found a warm home. The firemen had a red sweater made for the mascot. His collar was adorned with "H.F.D." and he had a badge. Blackie did not go on fire runs but hung around the station greeting visitors and soliciting a little scratching behind the ear.

FIREMEN, 2002, BLACK SHIFT. From left to right are (first row) Doc Knight, Tony Harris, Lee Jay Smith, Elmer Greene, John Burke, Mike Wilson, and Jeff Parker; (second row) Steve Masselli, Rodney Thomas, Ritchie Mabry, Rodney Croft, Brian Smiley, Gerald Morris, Eric Bleidt, and Gerald Higgins.

FIREMEN, 2002, GRAY SHIFT. From left to right are (first row) Tim Goodwin, Cecelia Cloos, Spencer Wood, Steve Futrell, Scott Wiesen, Tim McGar, Billy Palmer, and Donnie Miller; (second row) Glenn Pepper, Travis Lancaster, Steve Gallup, Irvin Cushenberry, Bill Campbell, Larry Turner, Tom Strack, Ken Calvin, and Eric Brown.

FIREMEN, 2002, WHITE SHIFT. From left to right are (first row) Mike Morris, Steve Futrell, Mike Noffsinger, Jay Carr, Scott Daniel, Edwin Moore, Mike Harris, Mike Henderson, and Alexandra Crook; (second row) Matt Wyatt, Scott Williams, Larry Hulsey, Steve Marquess, Pete Glass, Eddie Yokum, Mike Workman, Carl Long, Tim Fleming, and Howard Turner.

FIREMEN, 2004, GRAY SHIFT. From left to right are (first row) Steve Gallup, Larry Turner, James Upton, Chip Renshaw, and Brian Diamond; (second row) Kevin McDowell, Irvin Cushenberry, Bill Campbell, Dan Davis, Billy Palmer, Tim McGar, Glenn Pepper, Donnie Miller, Dennis Creamer, Tyler Askew, Matt Wyatt, Mike Pendleton, and Ken Calvin; (third row) Scott Bullard, Scott Wiesen, Bill Nightingale, Greg Boyd, Spencer Wood, and Tom Strack.

FIREMEN, 2004, WHITE SHIFT. Included in this picture are Alexandra Crook, Brent Sisk, Mike Henderson, Kevin Quarles, Jay Carr, Eddie Yokum, Mike Workman, Carl Long, Tim Fleming, Mike Morris, Jim Byrum, Gerald Higgins, Larry Hulsey, Steve Marquess, Steve Futrell, Freddie Montgomery, Greg Boyd, Keith Hampton, Curtis Nightingale, Tim Goodwin, Bill Pyle, Larry Dale Meacham, and Ron Luckey.

FIREMEN, 2004, BLACK SHIFT. Included in this picture are Johnny Ezell, Leejay Smith, Scott Peck, Kevin Quarles, Matt Bryant, Sam Ousley, John Burke, Brent Sisk, Carla Hill, Eric Brown, Doc Knight, Brian Smiley, Mike Wilson, Kurt Williams, Tyler Askew, Rodney Thomas, Spence Sowell, Donnie Cato, Greg Boyd, and Jeff Parker.

ANOTHER MASCOT. The members of the Hopkinsville Fire Department adopted a dalmatian in the spring of 1998. He was given the name Chief Concern and was provided with a red jacket complete with the seal of the fire department as a part of its decoration.

Firemen, 2007, Gray Shift. From left to right are (first row) Dennis Creamer, Jeff Slaughter, Spencer Wood, Bill Nightingale, Tim McGar, Billy Palmer, Spence Sowell, Jessica McQuade, and Brian Diamond; (second row) Larry Dale Meacham, Donnie Miller, Chip Renshaw, Larry Turner, Dan Davis, Kenny Calvin, Irvin Cushenberry, Keith Hampton, Locke Beasley, Mike Pendleton, John Burke, and Kirt Williams. Not pictured are Tom Strack, Glen Pepper, Scott Bullard, and James Upton.

Firemen, 2007, White Shift. From left to right are (first row) Troy Beach, Lucas Stagner, Scott Daniel, Jay Carr, Edwin Moore, Alex Crook, Andrew Watts, Kevin McDowell, and Mike Henderson; (second row) Larry Hulsey, Mike Workman, Curtis Nightingale, Steve Futrell, Ron Luckey, Jim Byrum, Tim Goodwin, Mike Morris, Bill Pyle, Gerald Higgins, and Freddie Montgomery. Not pictured are Carl Long, Matt Wyatt, and Eddie Yokum.

FIREMEN, 2007, BLACK SHIFT. From left to right are (first row) Rodney Thomas, Anthony Harris, Tommy McGraw, Kevin Quarles, Ryan Fugate, Eric Brown, Steve Gallup, Chad Sivils, and Matt Bryant; (second row) Carla Hill, Brent Sisk, Doc Knight, Scott Peck, Sam Ousley, Mike Harris, Johnny Ezell, Josh Buxton, Clay Walton, Jason Hale, Mike Wilson, Tyler Askew, and Jeff Parker. Not pictured are Ken Bybee and Mike Pentecost.

HONOR GUARD. The Hopkinsville Fire Department Honor Guard provides the opportunity for the department to be represented at special occasions. Memorial services and funerals of firemen offer the occasion to remember the sacrifices of those who strive to make our lives safer. The honor guard includes, from left to right, Rodney Croft, Elmer Greene, Troy Beach, Tony Harris, John Burke, and Harry Mason.

Toy Repair. The Hopkinsville Fire Department launched a tradition over 70 years ago that brings joy to less fortunate children at Christmas. Throughout this span of time, the firemen have conducted an annual toy drive, with toys brought to the fire station for the firemen to repair and distribute. This early 1950s image features Buster Wallace, at center, and other unidentified members of the force as they repair toys. This scene, from the East Ninth Street central fire station, features a side view of the 1928 American LaFrance pumper at left. Note the gas furnace at right, which replaced the coal-fired iron stove of an earlier era. The department also cut cedars, which were sold at the station for Christmas decoration. The money collected from these sales purchased needed items for community children.

Three

CHIEFS

FIRE CHIEF GEORGE E. RANDLE, 1901–1906. The first fire chief of the paid Hopkinsville Fire Department, George E. Randle (1861–1944), served the department for 22 years. He was a member of the volunteer department for 17 years and was chief for 5. He followed four other chiefs of the volunteer department, dating from the time fire records were first kept: H. D. McQuigg (1870–1871), Byron McQuigg (1871–1875), John H. Witty (1875–1880), and John A. Twyman (1880–1893). During Chief Randle's tenure, the department occupied the first new fire station, new horse-drawn equipment was purchased, extensive training programs were initiated, a town water system was installed, and a telephone fire-alarm system was placed in the new station. After retirement from the department, Randle operated a sheet-metal shop.

Fire Chief Ed H. Hester, 1906–1912. Edgar H. Hester (1871–1960) served as fire chief during a troublesome period. His department was confined under house arrest when the Night Riders, farmers upset at the low price they were receiving for their dark tobacco, raided Hopkinsville. Firemen were confined while the Night Riders set fires in two tobacco warehouses. Hester was a member of the force for 11 years. After his resignation, he spent the rest of his life in the contracting business.

Fire Chief Ed P. Fears, 1912–1918. Edward P. Fears (1861–1918) served the Hopkinsville Fire Department for 29 years. It was during his service that lightning struck and burned historic Cate's Mill on East Ninth Street at Little River. Fears died shortly after the end of his service.

Fire Chief Ernest S. Hayden, 1918–1930. Ernest H. Hayden (1872–1940) served as fire chief for 12 years but was an active member of the force for 30 years. Hopkinsville bought its first motorized fire trucks during his service as fire chief. Significant fires while Hayden was chief included the Central Fire Station in 1924, the Mogul Wagon Company in 1925, and Acme Mill and Planters Hardware, both in 1927.

Fire Chief Joe W. East, 1930–1952. Joe W. East (1885–1960) was the longest-serving chief in the history of the Hopkinsville Fire Department. He served on the force for 31 years, 22 as chief. The beloved leader of the department initiated a very successful fire prevention program. The fires at First Methodist Church and radio station WFIW, both in 1931, and the destruction of Hotel Latham in 1940 were the most significant in his tenure.

FIRE CHIEF DORIS E. JAMES, 1952–1958. Doris E. James (1912–1979) was a member of the fire department for 22 years. His tenure included the first purchase of new fire equipment in nine years and the Cayce Mill Supply fire in 1958.

FIRE CHIEF CARL T. MCGEE, 1958–1967. Carl T. McGee (1914–2006) served as fire chief for nine years. During his leadership, the city acquired its first new aerial ladder truck, and he annually led a very successful fire prevention program. In 2005, the retired fire chief was honored at the 100th anniversary of Hopkinsville High School Tiger Football as the oldest living player.

FIRE CHIEF FLOYD T. CANNON, 1967–1981. Floyd Cannon became a Hopkinsville firefighter in 1954 and served the department for 27 years, 14 as fire chief. The acquisition of new equipment—two new pumpers—was the highlight of his tenure. A commitment to continuing fire prevention programs was emphasized under his leadership.

ACTING FIRE CHIEF EARL E. CRAFT, 1981–1982. Earl Craft served as a member of the force for 26 years and filled the chief's position on an interim basis between Floyd Cannon and Paul Barnes.

Fire Chief Paul H. Barnes, 1982–1985. Paul Barnes focused on fire prevention and training during his years as fire chief. Also during Barnes's service, the city acquired a Ford Mini 300-gallon-per-minute pumper. This gave the firemen other outlets, with smaller capacity, for the department to use.

Fire Chief Jesse J. Lancaster Jr., 1986–1991. Jesse Lancaster's service as fire chief for five years included the purchase of three new pieces of equipment: an E-One Allegheny pumper, an air support truck, and a Pierce pumper. The expansion of equipment was very indicative of the necessity to better serve Hopkinsville.

FIRE CHIEF MIKE R. MAXWELL, 1991–1993. The last 10 years of the 20th century brought a changing trend in the tenure of fire chiefs. The rapid changes in technology and political influences shifted the emphasis from long-tenured chiefs to short periods of service. Mike R. Maxwell served as fire chief for two years and in his tenure bought a fire inspector's truck.

FIRE CHIEF WILLIAM F. JEFFRIES, 1993–2003. Bill Jeffries led the department in an era of acquisition of new heavy equipment: a Freightliner 50-foot aerial and pumper "Squirt" and the Quality 105-foot Quint aerial and ladder truck. These two pieces of equipment were bought at a cost of nearly $1 million.

ACTING FIRE CHIEF HARRY C. MASON, 2003. In the last half-century of Hopkinsville Fire Department history, "Daddy Harry" Mason has come to represent the spirit of the department. He joined the force in 1960 and served for 43 years. This tenure constitutes the longest period of service in the history of the department.

FIRE CHIEF FAGAN W. PACE JR., 2003–PRESENT. The current fire chief of Hopkinsville, Fagan Pace (seated), joined the department in 1987. He currently leads a department of over 80 firefighters who operate out of four stations with nine major pieces of firefighting equipment. Standing are deputy chief of training Eric Bleidt (left) and deputy chief of operations Brian Smiley.

Four

EQUIPMENT

SILSBY STEAM FIRE ENGINE. The oldest known piece of Hopkinsville fire equipment is this Silsby steam fire engine photographed in action by Christian County native Edgar Cayce on a Hopkinsville street around 1900. A hand-pump engine called the Captor was purchased by Hopkinsville in 1832. In 1837 and again in 1846, the town acquired hand-pump engines and hoses. In 1869, Silsby Peerless steam engine No. 202 was purchased. On February 29, 1884, the town council purchased a second Silsby steam engine, No. 771, for $4,000. A trade-in allowance of $1,000 was received for the older engine. The new engine fought large fires by dropping a suction house into one of many downtown fire cisterns, located at intersections. After the installation of a city water system in 1896, stronger city water pressure ended the need for this mechanical masterpiece. In September 1908, the engine was sold to Canton Pike truck gardener W. R. Brumfield, who used it to pump water out of Little River to irrigate his gardens. The old Silsby engine was sold in the World War II scrap-metal drives.

Horse-Drawn Ladder Wagon. In 1901, Hopkinsville firemen grouped around the ladder wagon for this image in front of the fire station on the southwest corner of Fifth and Main Streets. In addition to this wagon, the department had a chemical and hose wagon.

Ready for a Run. Fireman Joe East reins the horses from his seat on a wagon in this pose for the camera. It is the only known photograph of the interior of the Ninth Street central fire station prior to the 1924 fire that destroyed the building. The horse harness, suspended from the ceiling, was dropped into place over the horse's backs by means of a rope.

HORSES AND MEN. This 1910 scene features men holding the reins attached to four town celebrities. All fire horses were typically named; these are, from left to right, Dixie, Clipper, Snow, and George. This picture was taken in front of the old fire station on East Ninth Street. Policeman Abner W. Witherspoon stands at extreme left, and the short fireman is John Lawson.

POSED FOR THE CAMERA. Hopkinsville's new fire chemical and hose wagon is shown at the intersection of Ninth and Campbell Streets on April 28, 1909. It was purchased by the city and delivered on October 8, 1907. The firemen are John Hines at left and Fire Chief E. H. Hester at right. The fire horses are Snow and Dixie. The Latham cottages appear in the background.

MEN AND EQUIPMENT, 1930. This image shows the first motorized fire trucks purchased by the city of Hopkinsville in July 1919 and delivered by rail the following November. The trucks, bearing the Acme-Boyer brand name, were manufactured at Logansport, Indiana. The hose and chemical truck is featured on the right, and the ladder truck is located on the left. Pictured from left to right are Fred Tunks, Garland Morse, Bill Cannon, Pete Perry, Charlie Smith, and Herman Johnson. The trucks cost a total of $11,300.

TOP OF THE LINE, 1930. Hopkinsville firemen, from left to right, are Chief Ernest Hayden, Robert Boyd (on seat), Joe Morgan, Joe East (on seat), Guy Smith, Clarence King, Oscar Bates, Stanley Ward, John Lawson, and Austin Morgan. They are posing in front of the 1928 American LaFrance pumper at left and the 1927 Chevrolet pumper at right.

BUILDER'S PHOTOGRAPH. The Jaeger Portable Power Corporation of Detroit, Michigan, sold this 1927 Chevrolet 350-gallon-per-minute pumper to Hopkinsville for $3,050. Transferred to Riverside Cemetery for dirt hauling after World War II, it was sold in 1964 to Leland Davis. His son, Bob Davis of Little Rock, Arkansas, restored and gave it to the City of Hopkinsville in April 1999. It will soon be on display at the Woody Winfree Fire-Transportation Museum, located at the old East Ninth Street central fire station.

SOLD BY THE POUND. The City of Hopkinsville purchased this American LaFrance 1,000-gallon-per-minute pumper at one dollar per pound: $13,750. It was delivered by rail from Elmira, New York, on January 21, 1928. The American LaFrance was the first-run truck until 1940. The city sold the truck to John W. "Woody" Winfree on June 20, 1968, and Winfree presented it to the Pennyroyal Area Museum in 2004. The truck is currently being restored for inclusion in the museum.

Equipment and Men Line Up. This 1932 scene at the Ninth Street central fire station features, from left to right, the 1928 American LaFrance pumper, the 1927 Chevrolet pumper, and the 1919 Acme-Boyer ladder truck. The firemen include, from left to right, Chief Joe East, Charles Dulin, Ed East, Stanley Ward, John Lawson, Johnny Payne, Austin Morgan, Ernest Hayden, Pete Perry, and Garland Morse.

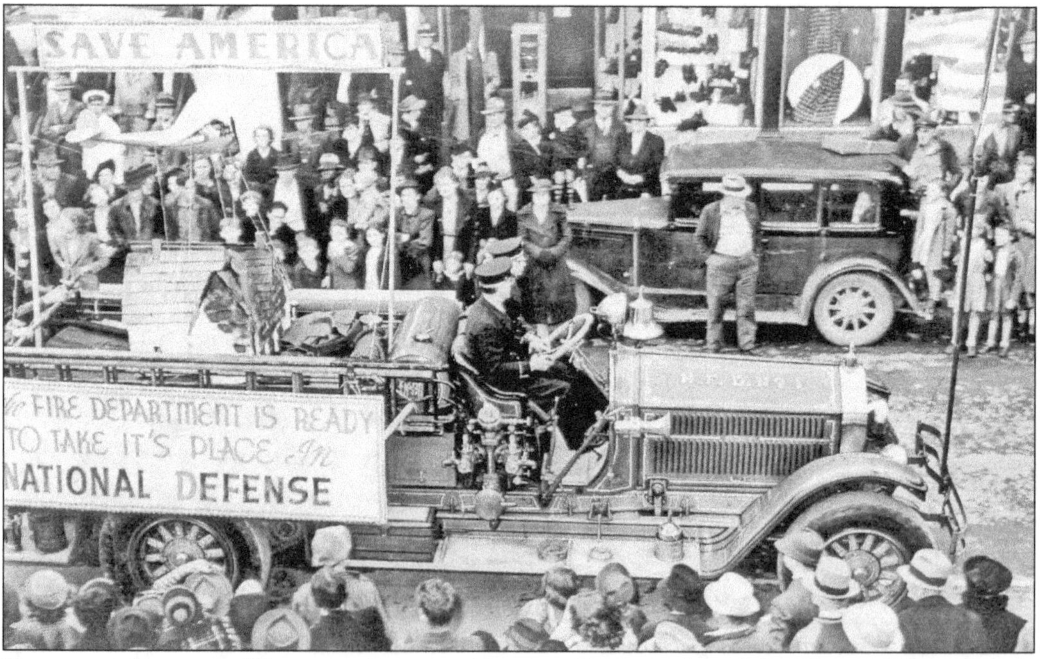

Ready for Defense. The scene shows an Armistice Day parade on Main Street, November 11, 1942. Firemen Ed East (driver's seat) and Charlie Dulin (shotgun) guide the big pumper through a large patriotic crowd. This photograph was made from an upstairs window in the Phoenix Building at Ninth and Main Streets.

POINT OF PRIDE. After 12 years without the acquisition of any new fire equipment, the city bought for $5,000 a new 1940 Dodge pumper, featured here in front of the Ninth Street Central Fire Station on August 18, 1947. The truck was delivered in December 1940, shortly after the Hotel Latham fire. It was retired from service, was sold to John W. "Woody" Winfree on January 3, 1979, for $750, and will become a part of the Woody Winfree Fire-Transportation Museum.

A NEW LADDER TRUCK. The department acquired a 1942 GMC truck in 1944 and after extension of the bed converted it into a ladder truck. It is pictured in front of the Ninth Street central station on August 18, 1947. After replacement by the new aerial truck in 1960, it was converted into a tobacco transfer truck.

NEW AND OLD. Hopkinsville firemen are photographed on line with the equipment in front of the Ninth Street central fire station. Trucks include, from left to right, a 1946 Dodge emergency van, the 1928 American LaFrance pumper, the 1940 Dodge pumper, and the 1942 GMC ladder truck. Firemen include, from left to right, Chief Joe East, Stanley Ward, Carl McGee, Buster Wallace, Ed East, Bill Griffin, Doris James, Joe Ward, Dennis Blandford, Haywood Scruggs, Forgey Mobley, Leland Davis, Pete Perry, and bystander Aubrey Wood.

A VOLUNTEER GROUP. In the spring of 1939, a volunteer fire department consisting of 15 members was organized at Western State Hospital, a mental institution with 1,900 patients. In 1953, the volunteer firemen are pictured demonstrating waterpower from a pre–World War II GMC water truck.

A New Pumper. In 1960, the city took delivery on a new Seagrave 1,000-gallon-per-minute pumper. This delivery photograph, made in front of the Ninth Street central fire station, includes (left to right) Fire Chief Carl McGee and city councilmen Dorris Chewning, K. Josh Mize, Dr. James H. Young, Luther Coward (on running board), Ira Thompson, and Austin Caudle. Mayor F. E. "Dutch" Lackey is shown standing behind the cab. The truck was sold in 1988.

First Aerial Truck. In 1960, the department acquired an 85-foot aerial American LaFrance, No. 103. City firemen had made repeated efforts to acquire such a truck as early as 1939, one year before Hotel Latham burned. The truck was sold to Lee Jay Smith.

Delivered by Rail. Fire chief Carl McGee stands on the running board as the 1964 American LaFrance pumper is unloaded from a railroad boxcar at a siding between Thirteenth and Fourteenth Streets. Labeled No. 7, it was later sold to the Gracey Volunteer Fire Department, where it is now in use.

Another Boxcar Delivery. On October 21, 1968, another new American LaFrance 1,250-gallon-per-minute pumper is unloaded from a railroad boxcar. Identified as No. 8, the truck cost $37,000 and was the first in Hopkinsville to have air brakes and power steering. The truck was sold in 1993.

HEADED FOR A NEW HOME. Hopkinsville State Farm Insurance agent John W. "Woody" Winfree placed the winning bid when the 1928 American LaFrance pumper was declared surplus and sold for $1,150. On June 20, 1968, fire memorabilia collector Winfree proudly drove this Hopkinsville fire truck home. After driving the relic in many parades and taking several children on enjoyable rides, Winfree donated the truck to the fire-transportation museum that now bears his name.

SECOND TIME AROUND. Hopkinsville fireman Walter Pruitt, the first African American on the force, points the way for John W. "Woody" Winfree to drive his second Hopkinsville fire truck home. The 1940 Dodge pumper was delivered to the city in December 1940. It was sold for $750 as city surplus on January 3, 1979, and it is now apart of the Woody Winfree Fire-Transportation Museum.

PRIDE OF 1964. In this 1991 photograph, Hopkinsville's No. 7, an American LaFrance pumper bought in 1964, departs for new ownership by the Gracey Volunteer Fire Department. It remains today as a vital part of the Gracey fleet.

EXPENSIVE FOR ITS DAY. This image was made as the 1968 American LaFrance pumper No. 8 was being sold in 1993. At a purchase price of $37,000, the truck was considered an expensive investment nearly 40 years ago.

THE FIRST DIESEL. In December 1975, the fire department took delivery on an American LaFrance 1,250-gallon-per-minute pumper. No. 9, as it was labeled, was the first diesel engine in the fleet and cost $54,900.

THE OLD WORKHORSE. The oldest member of the fire department's truck fleet is a Mack 1,250-gallon-per-minute diesel truck. The pumper-rescue vehicle, cataloged as No. 5, was put into service in February 1981. The truck is housed currently at Substation No. 4, located on Eagle Way Bypass.

SMALL WONDER. In 1983, the department received a Ford Mini 300-gallon-per-minute pumper. It was identified as No. 110 and was used extensively to service small house, car, and grass fires. This truck, like all other Hopkinsville trucks throughout the history of the fire department, bore red paint.

ANOTHER NEW PUMPER. A new E-One Allegheny 1,500-gallon-per-minute pumper was delivered to the city in 1987. No. 114, which is currently stationed at Skyline Drive Substation No. 2, was manufactured in Ocala, Florida. It was the first in local use to feature a four-man crew cab.

AIR SUPPORT. In November 1988, the department bought a new GMC 3500 air support truck. It bears the No. 106 and is housed at Substation No. 3 on the Canton Pike. This piece of equipment is used to provide extra air when fighting long-range fires. This unit was purchased for $27,000.

A NEW BRAND FOR HOPTOWN. The city received a new Pierce fire engine in 1991. No. 115 features a 1,500-gallon-per-minute pumping capability along with a 1,000-gallon booster tank. The closed cab is an outstanding feature in modern firefighting machinery construction, contrasted with the time when firemen rode in open cabs or hung on from the rear platform.

The "Squirt." A Freightliner FL80 built by the E-One Company in Nashville, Tennessee, was purchased by the department in January 1995. It is equipped with a 50-foot aerial ladder and a 1,250-gallon-per-minute pumper. Lettered in bold red and blue on the side of the ladder is the inscription "Hopkinsville Squirt, No. 107."

A Long Way from Steam Pumping. Engine No. 102, a Quality 1,250-gallon-per-minute pumper, was added to the department inventory in 1999. It represents one of the largest engines in the fleet and is currently housed at Substation No. 3, located on the Canton Pike.

PRIDE OF THE FLEET. In 2001, the department received its second new aerial truck. No. 108 is a Quality Gladiator FF 105-foot Quint aerial and ladder truck. It was purchased at a cost of $585,000 and is currently housed at Substation No. 4, located on the Eagle Way Bypass.

LET THERE BE LIGHT. A new addition to firefighting equipment in the fleet of the Hopkinsville Fire Department is an Allmand SHO lighting system. It was bought in 2005 at a purchase price of $14,000. The lighting system's four-light high-intensity output provides great assistance for nighttime firefighting.

LUXURIOUS RESCUE. The newest addition to Hopkinsville Fire Department equipment is a 2007 Chevrolet 1500 Suburban. This vehicle serves as a mobile command post and is located at the West First Street central station. This rescue vehicle was purchased by the department for $36,648.

NEW KID ON THE BLOCK. The year 2007 marked the arrival of a Chevrolet Polaris 6-by-6 Ranger. This all terrain brush vehicle, valued at $15,000, was a gift to the department by the United States Smokeless Tobacco Company, located in Hopkinsville. This vehicle is of great assistance in fighting field and brush fires, where it is difficult to gain entrance with a large pumper. Also pictured is the oldest piece of equipment in the department, a century-plus-old hose handcart used at Western State Hospital in the late 1800s.

Five

TRAINING

AERIAL TRAINING. A part of the mission of the Hopkinsville Fire Department and its predecessor volunteer companies is the training of firefighters. Throughout the 19th century, able-bodied men were expected to meet on a regular basis and practice firefighting with what meager equipment they had. Bucket brigades, hook-and-ladder companies, and engineers were required to be proficient and ready when the fire call came. The first training tower used by the Hopkinsville Fire Department was built in 1932 on the southwest corner of the lot owned by the local Masonic lodge east of the First Presbyterian Church. This scene, captured during World War II, illustrates the use of the training tower by members of the local firefighting force, while Fire Chief Joe East looks on from ground level. This training tower was removed about 1952. Note the ladder truck at left and the Dodge pumper on the right.

Training Demonstration. In 1950, the department built a new three-story, concrete-block training tower on the north side of Eighth Street between Liberty and Clay Streets, which is located behind the Ninth Street central fire station. In this scene, local firemen are joined by others from nearby towns to practice the raising of uncharged fire hose to the roof above the third-story level. Local firemen include Carl McGee (second from left), Buster Wallace (third from right), Forgy Mobley (lower end of the ladder), and Fire Chief Joe East (standing at left in the second-story doorway). This structure was utilized for training until the new five-story training tower was built in 1963–1964.

A New Tower. In 1963–1964, the new West First Street central fire station construction included the building of a five-story brick training tower. The structure stands on the bank of the north or town fork of Little River, directly behind the Central Station. The firemen in this photograph are engaged in the practice of lowering men by rope from the upper floors to the ground level.

All the Way to the Top. Fireman Bill Griffin, standing at left, watches the camera as members of the department ascend a ground ladder at left and the aerial ladder at center. This view, captured on September 30, 1964, shows training around the new tower. Equipment used in this operation included the 1960 American LaFrance aerial ladder truck, featured here. The men in this photograph, realizing a picture is being taken, showcase their strength and agility by climbing all the way to the top. Physical endurance is a vital part of firemen's training so they do not get fatigued in the field. As one fireman scotches the top, others climb to meet the assistance needs of their fellow colleagues. Notice the arms attached to the side of the fire truck. These arms provide balance and stability so that the firemen can complete their tasks.

HANGING FROM A THREAD. Fire Prevention Week has long been observed in October in Hopkinsville. Demonstrations of firemen's training have interested crowds for many years. In this October 1952 view, fireman Forgy Mobley is being lowered by rope from the roof of Charles Department Store, located on the northwest corner of Ninth and Main Streets. He is being assisted by local fire runner Woody Winfree.

A LOT GOING ON. This view from October 1949 features a part of the crowd and the equipment gathered for a surprise demonstration opening Fire Prevention Week. Mayor F. E. "Dutch" Lackey is shown standing in right foreground with a fire prevention program in his hands. Note city bus No. 8 at right, the 1927 Chevrolet pumper in center foreground, and from left to right across the street, the ladder truck, the 1940 Dodge pumper, and the 1947 Seagrave pumper.

AIR FOR THE INJURED. Following World War II, the local American Legion Post No. 233 presented this Dodge service vehicle to the Hopkinsville Fire Department for use as an emergency truck. Local firemen demonstrate the use of a portable oxygen case for a meeting of Western Kentucky firemen. Fireman Carl McGee is shown kneeling at the box, Leland Davis is pictured at left, and the victim is unknown.

PUMPING THE LAKE. This view features a demonstration of water being drafted at the water commission's lake at Hopkinsville Stone Company. Capt. Lonnie Webster, at right, explains the pump operation while Capt. J. E. Berryman, at left, demonstrates adjustment of fire streams. The 1947 Seagrave pumper is being used in this demonstration on April 26, 1955.

Men from Mars. An innovation came to local firefighting equipment with the addition of protective gasmasks. This demonstration was conducted by Jimmy Sholar (right) for two unidentified firemen at the East Ninth Street central station on July 28, 1961. Note the fireman's pole in the background and a map of the city of Hopkinsville on the wall at right.

High in the Sky. Familiar Hopkinsville landmarks provide a backdrop for a demonstration of the new 1960 American LaFrance aerial ladder truck. Local firemen have mounted the ladder above the tower of the First Presbyterian Church at Ninth and Liberty Streets in downtown Hopkinsville. In the background from left to right are the New Central Hotel, Standard Mill Supply Company, the East Ninth Street central fire station, and the town clock.

New Truck under Pumping Power. The new 1976 American LaFrance pumper No. 9 is getting a tryout on a cold February day in 1976. The 1,250-gallon-per-minute diesel pumper had just been delivered to the city of Hopkinsville at a cost of $54,900. The power of this pumper required two men on each hose.

Modern Training. These members of the Hopkinsville Fire Department practice the removal of victims from a fire-and-smoke-filled structure. This concrete building, located near the Hopkinsville–Christian County Airport, was used for fireman training through the 1990s. Observe the removal of a victim by lowering him to the ground with support from fellow firemen on the ladder.

Six

Fire Prevention

Ready for the Parade. In the days before television and computers, many community events and activities were centered on special occasions and celebrations. Parades and fire prevention awareness, along with school programs and demonstrations, were part of the Hopkinsville scene in October each year. The Hopkinsville Chamber of Commerce, the news media, civic clubs, and various churches strongly promoted this annual observance. The yearly event drew the community together in an effort to prevent fires. Fire chief Joe East, with the strong commitment of his department and the leadership of fireman Carl McGee, motivated various community organizations in a weeklong event of celebration and education. The 1942 GMC ladder truck, with a model of a frame house aboard, is about to enter the annual fire prevention parade. The East Ninth Street central fire station provides the backdrop for launching the event on September 28, 1947. The model house was constructed to be set on fire and extinguished in repeated demonstrations.

Boy Scout Orientation. A part of Fire Prevention Week was the distribution of fire literature in the residential sections of town. In this 1947 photograph, Fire Chief Joe East, standing in front of the East Ninth Street central station, discusses the project. Boy Scouts are, from left to right, Bobby Davis, Bobby Gerhart, Billy Sparks, Lewis Faulkner, Brooke Lawson, Billy Radford, Ed Bridges, Bobby Clint Cayce, and unidentified.

Girl Scout Participation. A local Girl Scout troop receives instructions from Fire Chief Joe East about passing out fire prevention literature at F. W. Woolworth Company and Montgomery Ward and Company. From left to right, they are Martha Helen Long, Mary Anne Garrott, Peggy Sue Wolfe, Annabelle West, Betty Duncan, Betty Hale, Mary Edith Stuart, Martha Ann King, Lois McGrew, Francis Owen, Patricia Todd, Beverly Boulter, unidentified, and Nita Wood.

FIREMEN'S SCHOOL. For a number of years, Hopkinsville was the site of the annual Western Kentucky Firemen's School. This two-day event attracted firemen from throughout Western Kentucky. This scene, captured on May 6, 1947, shows the crowd gathered in the lot across the street from the East Ninth Street central fire station. The photographer recorded the event from the old training tower.

BE AWARE. Fire Prevention Week activities included arranging downtown business show windows with the theme of stressing fire awareness. This unidentified show window in 1951 points out some of the dangers associated with home fires. Faulty wiring, wood stoves, and un-emptied trash cans were some of the dangers pointed out.

HERE COMES THE PARADE. Fire prevention parades, such as this one on Campbell Street, involve a number of community organizations'. This event in 1948 includes, among others, Beta Sigma Phi Sorority, Junior Chamber of Commerce, the Red Cross, and the Kiwanis and Rotary Clubs.

HIGHWAY SIGNS. This October 1952 scene points out the need for fire prevention 52 weeks in every year. Fire runner Woody Winfree stands at left, with Fire Chief Doris James looking on at right. They are standing on East Ninth Street near Little River opposite Chesnut and Anderson Grain Company.

ATTUCKS HIGH SCHOOL. Public school presentations and demonstrations were popular features of Fire Prevention Week. In this scene from 1950, Emmett Cox of the State Fire Marshall's Office speaks to students in the auditorium of Attucks High School. Note the size of the crowd gathered and the fact that the school library was located in a small room off the stage.

ACROSS TOWN. Norman Imrie of the State Fire Marshall's Office addresses the student body of Hopkinsville High School on fire safety in the first week of October 1949. The stage scenery backdrop was in place for a school play. Assembly programs such as this were weekly occurrences in the school located on Walnut Street.

ATTENTION CATCHER. In 1952, one of the feature attractions for raising interest in fire prevention was to parade "Gravey George" the dummy, pictured here with deputy fire marshal W. O. Spencer. He was placed on the 1928 American LaFrance pumper and paraded through the streets of Hopkinsville to make the public more aware of fire prevention.

SMOKEY THE BEAR. A feature attraction of one of the Fire Prevention Week's activities was Smokey the Bear. He is making an appearance among Boy Scouts at the front door of the Christian County Court House while Fire Chief Doris James looks on at right. Westside Elementary School principal G. Lloyd Seay is pictured at left.

MORE HANDOUTS. The 1947 Seagrave pumper and the First Presbyterian Church provide the background for fireman Carl McGee, pictured at left, and Boy Scout leader Bobby Clint Cayce at right. Fire Prevention literature is being given to these children by the two prominent individuals. Note the Barnes Motor Company used-car lot sign east of the church.

COOPERATION. Local businesses were honored with certificates of cooperation for their participation in Fire Prevention Week activities. In this 1950 scene, Fire Chief Joe East, at left, presents a certificate to New Central Hotel clerk J. D. Haywood. Ten years earlier, Haywood was the clerk on duty at Hotel Latham and reported the fire that destroyed the prominent landmark.

BANNER IN THE SKY. The annual Fire Prevention Week was promoted with a banner across a principle street in town. In September 1948, the banner was stretched across East Ninth Street from Higgins Drug Store on the left to the Phoenix Drug Store on the right. The town took this event seriously.

HOT WORK. The Hopkinsville Fire Department was committed to fire prevention in many ways. An annual and very hot job for the firemen occurred late each summer, as is revealed in this view of the back door of the East Ninth Street central fire station. Firemen recharge each of the fire extinguishers. The firemen collected, recharged, and returned the extinguishers at no charge.

The Message Rolls. Information provided for Fire Prevention Week was even advertised on large industrial trucks. From left to right, Capt. James H. Moore, D. C. Rockwood, and Fire Chief Carl McGee observe the message on a Thomas Industries truck. The Moelight Division of Thomas Industries manufactured lighting fixtures at its plant in Hopkinsville, located off North Drive.

Story Time. Smokey the Bear has put in an appearance in Mildred Baker's elementary class at Morningside School. This October 1964 scene shows the children's enthusiasm about a visit from such a well-known character. Note the desk style and aquarium at left.

LOCAL ADVERTISING. Local businesses supported the fire prevention movement through the distribution of calendars. Myers Alignment Center, located at Third and Liberty Streets, distributed copies of the official National Fire Protection Association calendars. In this October 1961 view, firemen James Nichols and James H. Moore show off local advertising.

MINSTREL SHOW. A popular entertainment emphasizing fire prevention was performed in minstrel shows. Fireman Carl McGee, on left, director of the local fire prevention program, is shown with Happy Kellam, the director of the minstrel show. Money raised from these events was used to promote the fire prevention effort in the community.

Seven

LANDMARK FIRES

COURTHOUSE RUINS, 1865. The summer after the Civil War was over, Hopkinsville's first photographer, Ezra L. Foulks, made this image of the ruins of the Christian County Court House. On December 12, 1864, Confederate general Hylan B. Lyon, with a small detachment, raided Hopkinsville and burned the courthouse along with the old Christian County Male Academy, which stood on the site of the Louisville and Nashville Railroad Freight Depot. General Lyon ordered the burning of the courthouse because Union men, suffering from the measles, had been quartered there. Volunteer firemen and private citizens formed a bucket brigade to keep the fire from spreading. Fortunately the county records, dating from March 1, 1797, were maintained in the Clerk's Office Building on the site of the present Alhambra Theatre. The 50-foot-by-75-foot, two-story courthouse, with a bell-equipped cupola, was constructed in 1837–1838 by John Buckner for $14,000. A 400-barrel-capacity cistern was dug on the back of the building for drinking water and fire protection. Rain water was guttered from the shingle roof to provide the water supply. This image is the oldest fire-related subject known to exist in Hopkinsville.

SECOND TIME AROUND. Thursday morning, November 2, 1905, the three-story brick South Kentucky College, located on Belmont Hill, was destroyed in a $20,000 fire. It originated from a defective flue in the attic. The fire started at 11:00 a.m. and burned until 1:00 p.m. Other school fires at this location occurred on February 24, 1884, February 2, 1912, and March 4, 1959, when Belmont Elementary School burned.

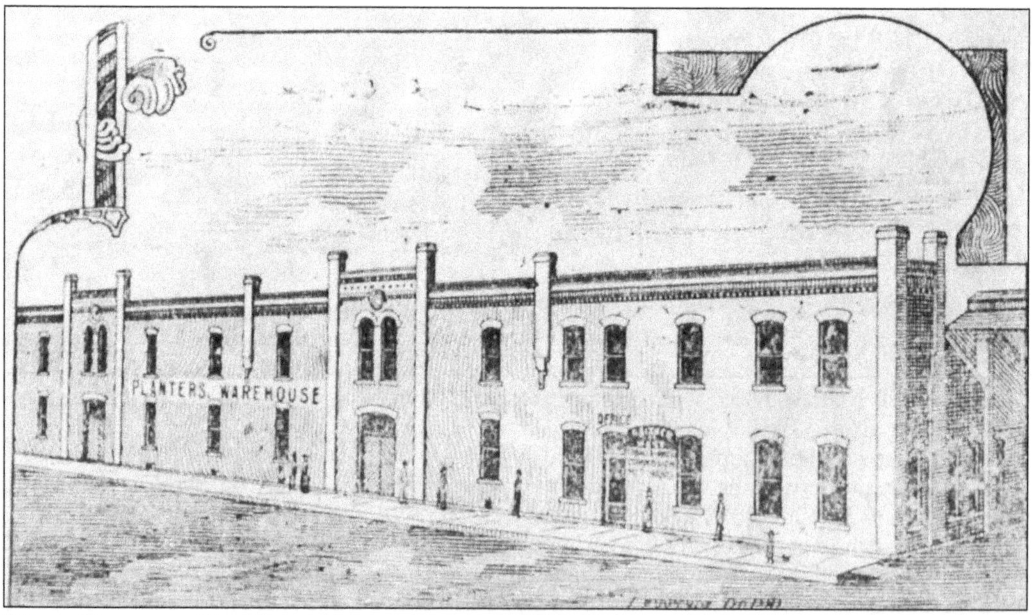

WHEN TOBACCO WAS KING. The production process for dark tobacco, the local No. 1 cash crop, included retail sales on the tobacco market at a warehouse. In 1889, Hopkinsville philanthropist John C. Latham Jr. built the Planters' Warehouse, located on the southwest corner of East Ninth and Campbell Streets. It was leased by M. H. Tandy and Company, who operated a tobacco sales floor at the location.

Ruins of the Planters' Warehouse. This image, looking southwest at the corner of Ninth and Campbell Streets, reveals the ruins of the Planters' Warehouse after a fire that took place on Saturday morning, December 7, 1907. Night Riders raided Hopkinsville and burned this warehouse, as it contained tobacco that belonged to farmers who would not join the Planters' Protection Association.

Another View. The Planters' Warehouse is shown in this photograph on Ninth Street after the pre-dawn raid by the Night Riders. In the background stand the ruins of the T. M. Wooldridge Tobacco Warehouse, which contained dark tobacco owned by the Night Riders and which caught fire from the one set at the Planters' Warehouse.

The Morning After. A small crowd of spectators gathers on the morning after the Night Riders raid to view the ruins of Tandy and Farleigh's Tobacco Warehouse. It was located on the southwest corner of Fourteenth and Campbell Streets and contained tobacco leaf owned by independent farmers who would not join the Planters' Protection Association. Acme Mills and Elevator Company is located in the center background.

Site of Peace Park. Ruins of two tobacco warehouses—T. M. Wooldridge on the left and M. H. Tandy on the right—give evidence of the physical layout of warehouses in present-day Peace Park. John C. Latham Jr., the owner of the block, bequeathed the land to the city of Hopkinsville when he died in 1909, with the direction that Peace Park be established on the site. The Louisville and Nashville Railroad Freight Depot appears in the center background.

CATE'S MILL DESTROYED. The milling facility of James Cate and Son lay in ruins when this picture was made a few days after its destruction. On September 2, 1918, lightning struck the roof of the grain elevator, setting the mill on fire. Within a few hours, the five-story frame structure was reduced to this scene. Mill products included wheat flour and cornmeal. A mill had been located at this site for over 125 years.

WAGON FACTORY FIRE. On Monday afternoon, December 28, 1925, the Mogul Wagon and Church Furniture Factory, located on the northeast corner of Twenty-first and Harrison Streets, was destroyed in a spectacular $500,000 fire. This view looking east toward the burning factory shows the frame office building in the center. Two hundred employees lost their jobs right after Christmas.

Fire Station Burns. As midnight approached on October 3, 1924, fire was discovered above the second floor of the East Ninth Street central fire station. Sleeping firemen on duty included John Lawson, Charles Dulin, Fire Chief Ernest Hayden, Joe East, Guy Smith, Otto Fish, and Joe Fish. Fireman Herman Johnson and his family were asleep in the apartment on the right side of the second floor.

Point of Embarrassment. Ruins of the stately structure reveal the East Ninth Street central station's total destruction down to the second floor. The two motorized fire trucks were quickly removed and later housed in Richard Leavell's livery stable next door to the east. The town clock tower and the historic fire bell collapsed at midnight, as the hour was being tolled. The loss was estimated at $15,000 with $6,000 insurance.

A Block in Ruins. The Planter's Hardware block, at center and right, and the four-story Pennyroyal Apartments at left, along with several other businesses, were destroyed in a $500,000 Sunday afternoon fire on November 6, 1927. In the back lot of the hardware building stood a tobacco warehouse filled with 7,500 bushels of sweet potatoes. Fire departments from Clarksville, Tennessee, and Earlington, Kentucky, came to assist in fighting the fire.

Fire in Lafayette. In the summer of 1931, a large fire destroyed several buildings, including stores and the post office in Lafayette, Kentucky, located 16 miles south of Hopkinsville. Hopkinsville sent the 1928 American LaFrance pumper to assist. On other occasions, the pumper was sent to aid in fighting fires at Gracey, Pembroke, and Crofton, Kentucky.

Owen Lumber Company, 1939. A $100,000 fire wiped out the Owen Lumber Company, wholesalers of hardwood lumber and flooring, on September 12, 1939. Located on Pardue Lane between the Lafayette and Cox Mill Roads, the plant had been in operation here since March 1936. Destroyed were three drying kilns, the engine room, the boiler room, the dust house, the tool house, the cooling shed, and thousands of feet of lumber.

Tragic Accident. On Thursday night, August 18, 1949, a tragic Clarksville Pike automobile accident, with death by fire of six Hopkinsville women, led eventually to the construction of a four-lane highway between Hopkinsville and Camp Campbell. A 1948 Dodge, in which the local women were riding, was struck from the rear by a 1941 Chevrolet containing three local soldiers. The impact forced the car into the saddle tank of a southbound car carrier containing five 1950 bullet-nose Studebakers.

Gant-Stites House, 1935. One of Hopkinsville's treasured landmark homes, located at 714 East Seventh Street, was constructed in 1851–1853 by Samuel L. Salter and John Orr from plans by Minard LaFever for William W. Western. It was bought by Joseph K. Gant in 1858, and after his daughter, Mary, married Dr. Frank M. Stites, the home was occupied by this family until it burned.

Ruins of the Stites House, 1945. In a $25,000 fire, the Stites House was destroyed on the morning of November 28, 1945. The fire originated in the basement and was fueled by the explosion of a gas hot-water heater. Nine people living in the landmark's four apartments narrowly escaped. Coauthor William T. Turner was a witness to this fire after having skipped his kindergarten class that morning.

CROFTON SCHOOL, 1948. Crofton Elementary and High School was constructed after its predecessor burned on May 26, 1924. The architectural style of this building is similar to the old Trenton High School. This structure served as the Crofton School for a little more than a quarter century. A gymnasium and a rear addition were built later.

CROFTON SCHOOL RUINS. On Saturday night, January 6, 1951, the old part and gymnasium of the Crofton School were destroyed in a $90,000 fire. A small fire was first detected in a supply room on the second floor. Arson was suspected, as an unlocked door was discovered early in the fire.

VIRGINIA STREET BAPTIST CHURCH. Virginia Street Baptist Church, located on the northwest corner of Third and Virginia Streets, was struck by a minor fire in the sanctuary in the 1950s. In the history of Hopkinsville, there have been very few church fires. The old Episcopal Church, located on the northeast corner of Fifth and Virginia Streets, burned on October 24, 1889. It was then being used as a storage building and icehouse after previous use as a grocery. A potentially dangerous fire was put out in the furnace room of First Baptist Church on November 27, 1912. The First Methodist Church sanctuary burned on March 12, 1931, and on February 16, 1948. Limited church fires in this community have been attributed to the outstanding involvement of local people in the fire prevention program. The only other church destruction locally has been by tornado when, on May 12, 1978, the steeples on the Second Baptist Church and the Grace Episcopal Church were demolished.

RECAPPING FIRE, 1955. On Friday night, November 24, 1955, a $50,000 fire destroyed the Hopkinsville Recapping Company. The business, located on the southeast corner of Thirteenth and Clay Streets, was owned by Carl M. Polley. The firm specialized in molding and tire vulcanizing work. The Fort Campell Fire Department assisted because of the air pollution created by the burning of tires.

JORDAN FURNITURE COMPANY FIRE, 1959. A $225,000 fire struck the Jordan Furniture Company on the night of February 25, 1959. Located on the southeast corner of Eighth and Main Streets, the fire originated in trash containers at the rear of the building. An earlier fire that evening was extinguished at the same location but apparently rekindled that night.

BELMONT SCHOOL, 1959. The site of Belmont Elementary School has been the location of four school fires, two during the day and two at night. At 12:30 a.m., on March 4, 1959, a fire probably caused by faulty wiring was discovered. A light breeze fanned the flames, which raged out of control until 3:30 a.m. Flames leaped high enough into the midnight sky to be visible for miles.

BELMONT RUINS. Coauthor William T. Turner snapped this photograph of the ruins of Belmont Elementary School a few days after the fire of March 4, 1959. The fire destroyed 14 classrooms, a gymnasium, an auditorium, and the cafeteria. They were all located in separate quarters. Area churches were used for classrooms until the new school was completed in 1960.

Towering Inferno. Drivers along Fort Campbell Boulevard experienced this spectacular view as they drove along in May 1966. The large barn and silo, long community landmarks, were located on the Old Clarksville Pike behind present-day Arby's restaurant and BB&T bank. The barn was utilized to house cattle, sheep, and horses.

Pal Lumber Company. On Saturday afternoon, August 2, 1969, a $500,000 fire destroyed the Pal Lumber Company. Owner Elton Lyle indicated the fire started from faulty wiring in the office, and it initiated a four-and-a-half-hour firefight. The lumber company was located off of North Drive and the Illinois Central Railroad.

LACY SCHOOL. As children were being delivered by bus to Lacy Elementary School on the morning of March 8, 1971, the school erupted in a fire that destroyed most of the landmark. An earlier school on that site had burned on January 7, 1945. Other county school fires have included Howell High School on November 24, 1939, and Pembroke High School on September 21, 1940.

SALUBRIA SPRINGS HOTEL. Salubria Springs Hotel, located two miles northwest of Pembroke, Kentucky, off U.S. Highway 41, was destroyed by fire in the late afternoon of December 17, 1976. The 22-room, two-story frame building was opened on June 20, 1908. This health resort was converted into the Christian County Benevolent Home in 1933. After use as a nursing home, it was filled with storage and hay, from which spontaneous combustion occurred.

Harvey House. The historic Harvey House, located at 1723 South Main Street, was struck by fire on April 23, 1980. The fire originated in the attic of the home and required the use of three pumper trucks and an aerial truck. It took five and a half hours to bring this fire under control.

Fully Engulfed. Firemen using a ladder make an assault on the second floor and attic area of the Harvey House on South Main Street. The home, built in the early 1890s, was occupied by the family of John Harvey and was owned by Jack and Gaynell Rittenberry at the time of the fire.

Grange Hall. The Church Hill Grange was organized December 2, 1873, out of a national organization called the Patrons of Husbandry, organized in Washington, D.C., in 1867. A farm family organization, it attempted to bring higher prices for farm produce and to create more sociability among rural people. The Church Hill Grange Hall was dedicated on November 22, 1878, and served as a meeting place for the group until it was disbanded in 1926.

Claimed by Fire. In the night of March 4, 1987, the Church Hill Grange Hall, located on the Cox Mill Road in the community of Church Hill, was destroyed by fire after a meeting of the Church Hill Community Club. The fire was caused by a defective flue from a coal-fired stove. Many records, photographs, and antique furnishings were destroyed.

The Thompson Building. Cornette's Office Supply, at 702 South Main Street, burned after 1:00 a.m. on December 31, 1984. It was located in the George O. Thompson Building, constructed in 1883. The photographic studio of Edgar Cayce, world famous psychic clairvoyant, was located on the second floor of this building from 1910 to 1912.

Funeral Home Fire. Around midnight, December 13, 2003, Gamble Funeral Home, located on the northwest corner of Fourteenth and Liberty Streets, erupted into flames. The fire, which nearly cost the owner, Raymond Gamble, his life, destroyed the funeral home before daybreak. The landmark was opened in 1954 as Henninger and Lilly Funeral Home and was later owned and operated by Sam Henninger.

Eight

HOTEL LATHAM FIRE

THE END OF AN ERA. On Sunday afternoon, August 4, 1940, beginning at 1:30 p.m., Hotel Latham was destroyed in a spectacular $250,000 fire. The fire originated in the attic, and the local fire department, aided by firemen from Clarksville, Tennessee, and Madisonville, Kentucky, fought a four-hour battle in an unsuccessful attempt to contain the fire. The hotel, located in the block on the east side of Virginia Street between Sixth and Seventh Streets, was designed by New York architect Edward S. Child. The 93-room hotel was built by the Hopkinsville Hotel Company in 1893–1894 and was opened on January 3, 1895. It was constructed in the Italian Renaissance style with a face of soft yellow brick and a red tile roof. The structure was named in honor of Hopkinsville native and Wall Street cotton brokerage banker John C. Latham Jr. Famous guests included Theodore Roosevelt, John Phillip Sousa and his band, Vice Pres. Charles W. Fairbanks, Ethel Barrymore, and Andy Devine.

Roof Involved. This view of Hotel Latham from the corner of Seventh and Virginia Streets indicates the progress of the fire by 2:00 p.m. The fire was reported about 30 minutes prior to the taking of this photograph. An onlooker observed smoke coming out of a large attic dormer window. The bridal suite is featured at the second-floor window level with awnings over each window.

Rear View. The camera lens caught this rear view of Hotel Latham from the intersection of Sixth and Liberty Streets looking southwest. Duffer Motor Company, the Dodge dealer, was located directly behind the hotel with a Standard Oil service station on the southwest corner. Grace Episcopal Church is just out of the picture at right. The elevator shaft and the kitchen chimney stack of the hotel are located at the center of the photograph.

A CLOSE UP REAR VIEW. About 30 minutes after the previous shot was taken, a closer view reveals the destruction of the elevator shaft, though the kitchen stack is still standing. The fire originated in the open attic above the fourth-floor rooms and was estimated to have been under way for about an hour before it was detected.

GAME CALLED OFF. Five minutes after the preceding picture, another photograph was taken along the Sixth Street side of the hotel. The Christian County Court House is in the right background. A Hoptown Hoppers Kitty League baseball game was under way at Mercer Park, located in the present West Ninth Street business section. The game was called off, and visiting team members were unable to retrieve their possessions.

FULLY INVOLVED. At 3:00 p.m., with two neighboring towns' fire departments sending pumper trucks to aid in the fight, the obvious objective was now to save the surrounding buildings. Another view from Sixth and Liberty Streets, with Grace Episcopal Church at right, shows a large crowd gathered to watch the death of a grand old tradition. Many in this Sunday afternoon crowd were equipped with cameras, and therefore over 100 pictures made that day have survived.

METAL CORNICE COLLAPSES. The elaborate metal cornice was the first debris to hit the sidewalks. The Seventh Street side of the building featured the ladies' entrance, and a red-and-white striped metal canopy extended from the door to the edge of the sidewalk. The southern exterior of the building faced the Greyhound bus station and the Cumberland Presbyterian Church.

A LOST CAUSE. A camera captured this image from Eighth and Virginia Streets at time the roof was fully involved. The roof of the Cumberland Presbyterian Church appears at right, and Kentucky Motor Company, the DeSoto Plymouth dealer owned by Cy M. Williamson Sr., is pictured in the center.

A HARD PULL. The furnace chimney rises above the burning hotel, with Duffer Motor Company at right. This scene was taken looking west along Seventh Street from Liberty Street toward Virginia Street. Firemen attempt to raise a hose to the roof of the automobile garage located across the alley from the hotel.

CHEVROLET PUMPER AT WORK. The intensity of the fire and the heat buildup required moving all of the fire trucks involved away from the site. The 1927 Chevrolet pumper was moved to a fire plug located at the intersection of Eighth and Virginia Streets. This view of the fire, also featured on the cover, shows the swelling of the crowd.

MILLING AROUND. As word spread about the fire, a larger crowd assembled to watch the action, as it was entertainment for the time period. This view at Eighth and Virginia Streets looking north provides an observation of the magnitude of the fire. During this time, many civilians were at work removing contents from the hotel.

GIVING DIRECTION. One of the firemen, walking near the 1927 Chevrolet pumper, appears to be giving direction to fireman Carl McGee at the front of the truck. Phillips 66 gives an example of petroleum signage in that era, nearly 70 years ago. Also note the intersection street light that hangs high overhead on the left.

GREETINGS. The main entrance, located on Virginia Street, with its red-and-white metal awning, provided a warm and friendly greeting to many hotel guests. Throughout its 45-year life, Hotel Latham was the scene of many elaborate banquets, civic club dinners, political gatherings, weddings, honeymoons, and even a funeral. Smoke billows from the windows of the Grand Parlor, located on the second floor.

Rooftop View. Photographer Roy A. Stadelman climbed to roof of the Postel Building, located on the northwest corner of Sixth and Virginia Streets, to get this shot of the dying hotel. About 4:00 p.m., the chimney stacks stand, though all would fall in a few minutes.

Old Veteran at Work. The 1919 Acme Boyer chemical and hose truck, with the deluge gun mounted on its side, works valiantly pumping water onto the blaze. At this point, it was considered a lost cause. This scene features the hotel from the corner of Sixth and Virginia Streets.

A Useless Fire Escape. This view at the corner of Sixth and Virginia Streets, similar to the previous one, provides a sweep of the Sixth Street side of the hotel. Fire escapes had been added to the structure during the remodeling of the building in 1912. Frank Boyd's Barbershop was located in the corner first-floor store room for many years. As the fire progressed, the fire escapes would twist and fall next to the barbershop.

Late Afternoon. This scene, looking west along Sixth Street, portrays the building in the last stages of the fire. The Dodge automobile garage, located behind the hotel, had been emptied of new and used cars. Little was left to be done except to keep the fire from spreading.

Same Action from Opposite Directions. These two scenes were photographed at almost exactly the same moment. The view above, looking east, catches the early stage of collapse of the chimneystack nearest Virginia Street. The 1919 Acme Boyer chemical and hose truck is in view at right. The view on the bottom reveals the same stack falling, though the camera is aimed westward along Sixth Street toward the courthouse in the background. At this point in both photographs, the entire building had been consumed by fire. Notice the smoke now coming from the first floor in the lower image.

The Front Wall Collapses. These two scenes, the before and during action shots of the collapse of a portion of the front wall, were made from opposite directions. The upper view looks south on Virginia Street, while the shot taken showing the wall's collapse was made looking north. Firemen in both pictures appeared to anticipate the coming collapse. After the fire, several days passed before the surrounding streets were cleared of debris. Many of the snapshots of the hotel fire were made by amateur photographers. The best photographs of the event were made by Roy A. Stadelman, who made the lower image on this page. Herdy Miller came to the fire that afternoon equipped with an 8-millimeter home movie camera loaded with color film. His five-minute film, consisting of action shots made throughout the afternoon, is a classic piece of filming.

CHIEF PROVIDES A HAND. The Virginia Street front of the building reveals the final stage of the fire. The chief of the Clarksville Fire Department is assisted by civilians in controlling a charged hose to combat the fire at the corner of Seventh and Virginia Streets. Faxon's Drug Store was located on this corner.

VIEW FROM EIGHTH AND VIRGINIA STREETS. The late afternoon witnessed the almost total destruction of Hotel Latham. This view from Eighth and Virginia Streets indicates that a large crowd was gathering to be a part of the event. Large burning embers, called firebrands, landed on the roofs of several downtown buildings, starting at least 14 fires.

Flying Debris. Late in the afternoon, another camera shot featured the south side of the hotel extending along Seventh Street. Firemen continue to wrestle with the fire hose as they attack the blaze from the back of the building. A portion of the cornice is seen flying through the air to the ground below.

Smoke from the Kitchen Stack. A camera lens caught this view of Hotel Latham from the intersection from Seventh and Liberty Streets. As water is being poured onto the south wall along Seventh Street, the wall and chimneystacks collapse into the street. Note the white smoke pouring from the kitchen chimneystack.

IT'S OVER. These photographs, made from opposite directions, reveal the work of the fire's destruction by late afternoon. The top image was made looking west on Seventh Street; the interior of the building is shown after the wall finally collapsed. The bottom picture, taken from the east, gives a different perspective in the moments just before the large chimney furnace at right collapsed through the roof of the automobile garage to its right. The repair to the roof in the garage was recognizable until the building was torn down in 2006. The ruins smoldered for several days after the fire, and the effort to clear the lot, led by Dave Young using a small Ford tractor and a pond scraper, took about three months. Laborers worked for months cleaning the brick, and two buildings, a farm machinery garage and a freight line service warehouse and loading dock, were constructed using the brick.

Nine

Rural Volunteer Fire Departments

Robert "Pop" Cook. Volunteer fire departments have been in operation in growing numbers around Christian County for over 40 years. Between 1962 and 1997, 14 county volunteer departments were organized. As of 2007, these 14 departments collectively have 32 fire pumpers, 20 tankers, and 12 brush trucks that provide fire protection for the second largest county in the state. There are currently 200 men and 40 women who serve as volunteer firefighters. The establishment of a countywide water system was essential to provide adequate fire protection. In 1977, the first county residential water hook-on occurred at Empire, north of Crofton. On September 23, 1980, the Christian County Firefighters Association was organized to contribute mutual aid for better fire and rescue protection throughout the county. The president of this organization and sole charter member was Robert "Pop" Cook, then director of the Christian County Disaster and Emergency Services. In this 1980 photograph, Pop Cook gives signals for a fire truck to douse a blaze during fires along the Louisville and Nashville Railroad tracks.

CROFTON VOLUNTEER FIRE DEPARTMENT. The Crofton Volunteer Fire Department serves a large section of north Christian County. Crofton is located 13 miles north of Hopkinsville on U.S. Highway 41. The department was organized in 1962 and has 18 firefighters currently on the force. The main station was constructed in 1984, and the substation was constructed in 1992. Each is a two-bay station located on opposite sides of the CSX railroad in Crofton. This arrangement makes possible the fighting of fires on either side if a freight train is stopped on the siding blocking the crossings. Featured above is a post–World War II Chevrolet pumper and ladder truck, the first in use by the department. Below, a lineup of the trucks and some of the volunteer firefighters stand in front of the substation, located on the east side of the track.

Lafayette Volunteer Fire Department. Lafayette Volunteer Fire Department, located in the town of Lafayette, 16 miles south of Hopkinsville, was originally organized in the 1950s by Lafayette merchant James W. "Jimmy" Lander. The picture above shows the International pumper that came to the department from the Civil Defense program and was promoted by county officials. Men include, from left to right, Burgess Combs, disaster coordinator; Malcolm "Buggs" Pritchett; Edgar Gee; and county judge William F. "Bill" Edmunds. The Lafayette Volunteer Fire Department was reorganized in 1981 out of the South Christian Fire Department. The current three-bay station, pictured below, was constructed in 2007. There are 24 firefighters, a pumper, a tanker, and a brush truck in this organization.

HIGHLAND VOLUNTEER FIRE DEPARTMENT. The Highland Volunteer Fire Department is located on the northeast edge of Hopkinsville near the former Highland Elementary School and Cave Spring Cemetery. The department was organized in 1969, when a one-bay building was constructed. Two more bays were added in 1974. The department has 15 firefighters, three pumpers, two tankers, and one brush truck. The photograph above shows the original one-bay structure as work is under way on the 1974 addition. The lower image features, from left to right, the 1967 American LaFrance pumper, one of only two yellow fire trucks in the entire county; the 1976 tanker; and the 1998 Chevrolet tanker.

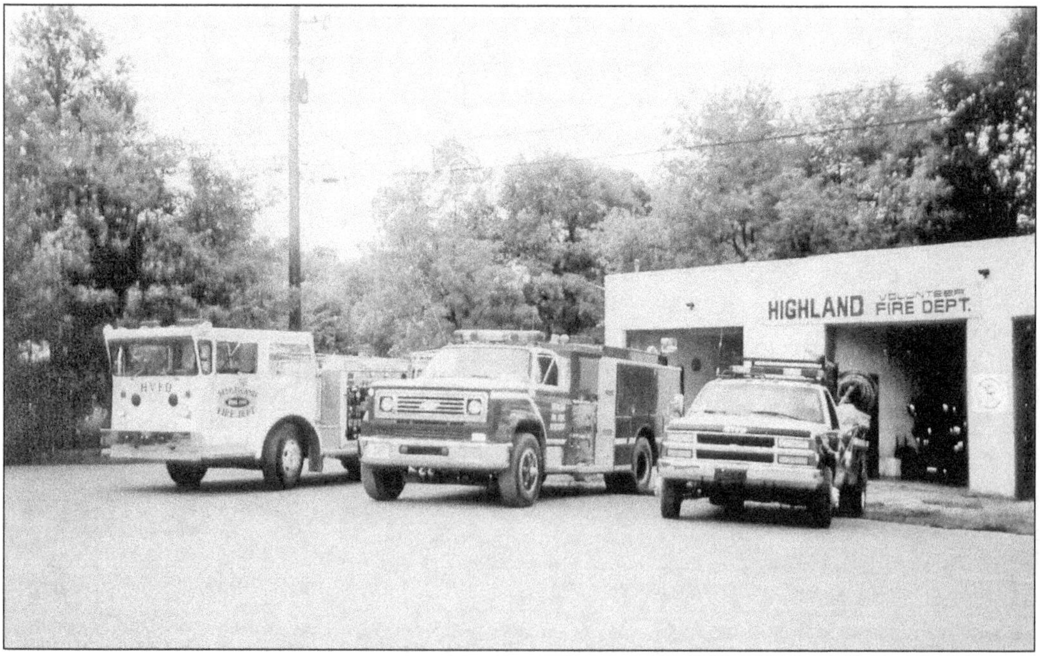

PEMBROKE VOLUNTEER FIRE DEPARTMENT. The Pembroke Volunteer Fire Department is located in the small community of Pembroke, Kentucky, located 10 miles southeast of Hopkinsville. The department was organized in 1970 and has 24 active volunteer firefighters. The current four-bay station was constructed in 1986–1987. The image above shows the 1962 International hose truck and volunteers (left to right) Dick Simmons, Terry Gee, Alton Slaughter, and Bill Rives. The photograph below features, from left to right, the 1976 Chevrolet tanker, the 1973 Ford tanker, the 1984 Ford high-pressure fog tanker owned by the city of Pembroke and only used within the city limits, and a 1979 Ford pumper used throughout the fire district. Note the hand-pull hose cart similar to the one at the West First Street central fire station located in Hopkinsville.

Oak Grove Volunteer Fire Department. The Oak Grove Volunteer Fire Department is located at Oak Grove on U.S. Highway 41A east of Fort Campbell, Kentucky. The department was organized in September 1974 and has 14 active volunteer firefighters. The five-bay station was constructed in 2000. The picture above features the 1979 GMC water tender, the oldest piece of equipment in the department. The lower image features the station and trucks, including, from left to right, the 2002 Freightliner pumper, a 1993 International pumper, the 2001 Ford fast attack/rescue pumper, a 1986 GMC pumper in yellow, and a 2006 Ford MED 01 ambulance.

West Side Volunteer Fire Department. The West Side Volunteer Fire Department is located off the Dawson Springs Road near the northwest bypass. The department was organized in 1977 and has 15 active firefighters on its force. The current four-bay station was built in 1999. The upper picture features the first new fire truck, a Ford pumper. Also visible in the image is construction of the first station, built in 1977 on the Dawson Springs Road. The lower picture, featuring the new station and equipment, includes, from left to right, a 1980 Ford brush truck; a 1991 Ford tanker; a 1986 Spartan pumper; and a 1982 Mack pumper, the first in the county with the color scheme of black and white.

LACY VOLUNTEER FIRE DEPARTMENT. The Lacy Volunteer Fire Department is located on Highway 107 North at Fearsville, Kentucky. It was organized in the spring of 1976 and currently has nine active firefighters. The current station was built in 1977. The image above shows the new three-bay station before a fourth bay was added in the 1990s. The lower photograph features the equipment in 2007; from left to right are a 1979 GMC tanker, a 2001 International pumper, the 1978 Chevrolet brush truck, a 1978 Chevrolet mini-pumper, and a 1970 Ford tanker. The Lacy Fire Department, like all of the others, annually conducts barbecues, fish fries, potluck suppers, and musical entertainment to raise money for department needs.

HERNDON VOLUNTEER FIRE DEPARTMENT. The Herndon Volunteer Fire Department is located in Herndon, Kentucky, on Highway 107 South, 10 miles from Hopkinsville. The department was established in June 1977. The department has 18 active volunteer firemen. The first three-bay building was constructed in 1978. The second two-bay building, with a meeting room and fully equipped kitchen, was constructed in 1994. The image above shows Herndon's first fire truck, a 1950 Chevrolet 500-gallon-per-minute pumper bought in 1977 for $2,000. After many years of service, it was sold for $2,500. The photograph below features two trucks in Herndon's fleet, a 1985 Mack pumper-tanker (left) and a 1989 E-One pumper (right).

GRACEY VOLUNTEER FIRE DEPARTMENT. Gracey Volunteer Fire Department is located in Gracey, Kentucky, on U.S. Highway 68, ten miles west of Hopkinsville. The department was organized in October 1977 and has 15 active volunteer firefighters. The first two-bay station was constructed in 1977, and the new three-bay station was built in 1992. The picture above shows the current department chief, Wayne Yates, sitting on the fender of a 1974 2500 Army brush truck. The scene below includes, from left to right, a 1992 tanker-pumper, the 1974 2500 Army brush truck, a 1964 American LaFrance pumper formerly owned by the Hopkinsville Fire Department, and a 1980 Chevrolet mini-pumper.

Mannington Volunteer Fire Department. Mannington Volunteer Fire Department is located in Mannington, Kentucky, north of Crofton near the Hopkins County line. The department was established in 1977. It maintains 30 active volunteer firefighters. The original two-bay station was constructed in 1978–1979, and a two-bay addition was built in 2001. The image above features the original trucks and a large group of department members. The 2007 photograph below includes, from left to right, a 1992 Plymouth medical and command van, a 1985 Ford brush truck, a 1977 Ford pumper, and a 1965 Ford tanker.

SQUARE DEAL VOLUNTEER FIRE DEPARTMENT. Square Deal Volunteer Fire Department is located on Highway 272 west of Hopkinsville. The department has 14 volunteer firemen and was organized in August 1978. The four-bay station was constructed in 1979–1980. The photograph above includes volunteer firemen, from left to right, (first row) Melvin Brown, Jeff Pryor, Belmont Sholar, Keith Sholar, Fire Chief Kenneth Hancock, Gene Shanklin, Edwin Hancock, Bobby Brown, and Jose Brown; (second row) Ken Hancock, John Boyd, and Roy Merritt. The lower image shows, from left to right, a 1979 GMC mini-pumper, a 2005 Ford pumper, a 1987 Ford pumper-tanker, a 1966 Ford tanker, and a 1970 Army brush truck.

FAIRVIEW VOLUNTEER FIRE DEPARTMENT. Fairview Volunteer Fire Department is located in Fairview on U.S. Highway 69, ten miles east of Hopkinsville. The department was organized in September 1978 and has six active volunteer firefighters. The three-bay station was built around 1985. The top image features the oldest piece of equipment in the department, a 1961 Ford pumper. The lower image shows the fleet in 2007; from left to right are a 1972 Jeep brush truck painted yellow, a 1976 Ford tanker, and the 1961 Ford pumper. Fairview is one of the oldest settlements in Christian County and is the birthplace of the only president of the Confederate States of America, Jefferson Davis.

NORTHWEST VOLUNTEER FIRE DEPARTMENT. Northwest Volunteer Fire Department is located on Highway 800 west of Crofton, Kentucky, and near Highway 109 at the entrance to Pennyrile State Park. The department was organized in 1986 and has 18 active volunteer firefighters. The first three-bay station was built in 1987 with a kitchen and meeting room added in 1993. A new five-bay station was built in 2006. The image above shows the original building constructed by the department when it was organized. The lower photograph features the new five-bay structure with, from left to right, a GMC mini-pumper, a John Deere utility vehicle, a 2003 Freightliner pumper with a Mercedes Benz power train, a 1967 Mack pumper, and a 1966 Ford tanker.

Honey Grove Volunteer Fire Department. Honey Grove Volunteer Fire Department is located 13 miles northeast of Hopkinsville on Highway 508. It was established in November 1996 and has 15 active volunteer firefighters. The first two-bay building was built in 1992, when the group formed a substation for Fairview Volunteer Department. The second three-bay building was constructed in 1999. Equipment includes, from left to right, a 1999 Kenworth tanker, a 1965 Ford pumper, and a 1977 2500 tanker.

Christian County Rescue Squad. The Christian County Rescue Squad, located on Sanderson Drive on the north side of Hopkinsville, was organized in 1971 by Earl Craft, Henry Gillette, and Robert Moody, who served as chief for 20 years. There are 33 certified rescue team members today. The building was occupied in 1980. Some of the equipment includes, from left to right, a 1993 Chevrolet Suburban Amhearst, a 2004 Ford rescue truck, and 1995 Ford rescue truck.